Letts
D0550402
revise GCSE
Bob McDuell
Chemistry

Contents

Further carbon chemistry

Further quantitative chemistry

Electrochemistry and electrolysis

Collection of gases

11 Food and drugs

Radicals

Group 2 metals

This book and your GCSE course

Awarding Body	AQA	EDEXCEL A	EDEXCEL B
Web address	www.aqa.org.uk	www.edexcel.org.uk	
Syllabus number	3421	1530	1539
Modular tests	None	None	6 tests each of 20 mins 30%
Terminal papers	1 paper. 135 mins 80%	Paper 1, $1\frac{1}{2}$ hours. Paper 2 (Extension) 1 hour.	3×30 mins. 2 papers on core, 1 on extension. 50%
Coursework	20%	20%	20%
Core Chemistry			
1. Classifying materials	10.1–10.2	C1	3 and 10
2. Changing materials	11.3, 10.5–10.6, 10.8–10.12	C2 and C4 and C6	4 and 9
3. Patterns of behaviour	10.13–10.14, 10.17–10.20	C3 and C5	4, 9 and 10
Extension Chemistry			
4. Water			
Hardness of water	10.15		
Soaps and detergents			
Solubility	10.15		
Purification of water supplies	10.15		
5. Acids, bases and salts			
Acids and neutralisation	10.15		
Salt production	10.15	C7	15
Tests for ions	10.16	C7	15
6. Metals and redox			
Lack of reactivity of aluminium	10.7	C8	16
Redox			
Rusting and its prevention			
Cell and fuel cells			
Alloys	10.7	C8	16
Steel manufacture	10.7	C8	16
7. Carbon chemistry	see note 1.		
Hardening oils	10.4		
Structures and isomers	10.4	C8	16
Ethanol	10.4	C8	16
Ethanoic acid and esters	10.4	C8	16
Thermosetting and thermoplastic	10.4		
8. Quantitative			
Simplest formula from percentages			
Mole	10.15	C7	15
Volume changes from equations		C7	15
Concentration in mol/dm³	10.15	C7	15
Acid–alkali titrations	10.15	C7	15
9. Electrochemistry and electrolysis			
Examples of electrolysis	10.7		
Quantitative electrolysis			
10. Collection and testing gases			
Collection and tests for common gases		C7	15
11. Food			
Chemistry of food and drugs			
12. Radical chemistry			
Radicals and ozone layer			
13. Group 2 metals			
Study of group 2			

Extension includes tests for carbon–carbon bonds and combustion of hydrocarbons 2.1

Visit your awarding body for full details of your course or download your complete GCSE specifications.

STAY YOUR COURSE!

Use these pages to get to know your course

- *Make sure you know your exam board*
- *Check which specification you are doing*
- *Know how your course is assessed:*
  - *– what format are the papers?*
  - *– how is coursework assessed?*
  - *– how many papers?*

OCR A option A	OCR A option B	WJEC	NICCEA
www.ocr.org.uk		www.wjec.co.uk	www.ccea.org.uk
1981		125	
None		None	None
90 min paper on Core, 45 min on Core and Extension. 80%		Foundation 2 hours. Higher $2\frac{1}{2}$ hours.	Foundation Paper 1, 1 hour. Paper 2, $1\frac{1}{2}$ hours. Higher Paper 1, $1\frac{1}{2}$ hours. Paper 2, 2 hours.
20%		20%	25%
3.1, 3.3, 3.5		C1, C4	3.1
3.1, 3.2, 3.5, 3.6, 3.7, 3.8		C3, C7, C8, C9, C10, C11	3.2, 3.5, 3.7, 3.8, 3.9, 3.13, 3.14
3.1, 3.3, 3.4, 3.5		C2, C3, C5, C6, C10	3.7, 3.11, 3.12
A1	B1	12	3.15
	B1		
A1	B1		3.15
	B1	12	
A2		5	3.6
A2		5	3.6
A2		3	3.6, 3.14
		9	
A3			3.7, 3.8
A3		9	3.7
	B3		
A3			
	B2		
A4		11	3.16
A4		11	3.16
A4			3.16
		11	
A5			3.10
A5		8	3.10
A5			3.10
A5		8	3.10
A5		8	3.10
			3.2
A6	B3		3.2
A6			
		2 (also chlorine)	Solids, liquids and gases (3.3) and Elements, mixtures and compounds (3.4) are KS3 in England and Wales
	B2	Plate tectonics is included in WJEC Chemistry	
	B3		

EDEXCEL A AND B, WJEC and AQA include manufacture of sulphuric acid (3.4)

Preparing for the examination

Planning your study

The final three months before taking your GCSE examination are very important in achieving your best grade. However, the success can be assisted by an organised approach throughout the course.

- After completing a topic in school or college, go through the topic again in Letts GCSE Chemistry Guide. Copy out the main points again on a sheet of paper or use a highlighter pen to emphasise them.
- A couple of days later try to write out these key points from memory. Check differences between what you wrote originally and what you wrote later.
- If you have written your notes on a piece of paper, keep this for revision later.
- Try some questions in the book and check your answers.
- Decide whether you have fully mastered the topic and write down any weaknesses you think you have.

Preparing a revision programme

At least three months before the final examination go through the list of topics in your Examination Board's specification. Identify which topics you feel you need to concentrate on. It is a temptation at this time to spend valuable revision time on the things you already know and can do. It makes you feel good but does not move you forward.

When you feel you have mastered all the topics spend time trying past questions. Each time check your answers with the answers given. In the final couple of weeks go back to your summary sheets (or highlighting in the book).

How this book will help you

Letts GCSE Chemistry Guide will help you because:

- it contains the essential content for your GCSE course without the extra material that will not be examined
- it contains Progress checks and GCSE questions to help you to confirm your understanding
- it gives sample GCSE questions with answers and advice from an examiner on how to improve
- examination questions from 2003 are different from those in 2002 or 2001. Trying past questions will not help you when answering some parts of the questions in 2003. The questions in this book have been written by experienced examiners who are writing the questions for 2003 and beyond
- the summary table will give you a quick reference to the requirements for your examination
- marginal comments and highlighted key points will draw to your attention important things you might otherwise miss.

Five ways to improve your grade

1. Read the question carefully

Many students fail to answer the actual question set. Perhaps they misread the question or answer a similar question they have seen before. Read the question once right through and then again more slowly. Some students underline or highlight key words in the question as they read it through. Questions at GCSE contain a lot of information. You should be concerned if you are not using the information in your answer.

2. Give enough detail

If a part of a question is worth three marks you should make at least three separate points. Be careful that you do not make the same point three times. Approximately 25% of the marks on your final examination papers are awarded for questions requiring longer answers.

3. Quality of Written Communication (QWC)

From 2003 some marks on GCSE papers are given for the quality of your written communication. This includes correct sentence structures, correct sequencing of events and use of scientific words.
Read your answer through slowly before moving on to the next part.

4. Correct use of scientific language

There is important Scientific vocabulary you should use. Try to use the correct scientific terms in your answers and spell them correctly. The way scientific language is used is often a difference between successful and unsuccessful students. As you revise make a list of scientific terms you meet and check that you understand the meanings of these words.

5. Show your working

All Chemistry papers include calculations. You should always show your working in full. Then, if you make an arithmetical mistake, you may still receive marks for correct science. Check that your answer is given to the correct number of significant figures and give the correct unit.

Core material

Topic	Section	Studied in class	Revised	Practice questions
1.1 Atomic structure	Particles in an atom			
	Atomic number and mass number			
	Isotopes			
	Arrangement of electrons in an atom			
	Link between reactivity and electron arrangement			
1.2 Bonding	Ionic (or electrovalent) bonding			
	Covalent bonding			
	Metallic bonding			
	Giant and molecular structures			
	Bonding, structure and properties			
	Allotropy			
2.1 Chemicals from organic sources	Refining crude oil			
	Burning alkanes			
	Making addition polymers			
	Uses of addition polymers			
2.2 Useful products from metal ores and rocks	Products made from rocks			
	Extracting metals from ores			
	Extracting metals by reduction			
	Purifying metals by electrolysis			
	Extracting metals by electrolysis			
2.3 Useful products from air	Ammonia			
	Nitrogen fertilisers			
2.4 Quantitative chemistry	Equations			
	Relative atomic mass and relative atomic formula mass			
	Using equations to calculate masses			
	Working out chemical formulae			
2.5 Earth cycles	Changes in composition of atmosphere and oceans			
	Carbon cycle			
	The rock record			
3.1 The periodic table	Structure of the periodic table			
	Development of the periodic table			
	Relationship between electron arrangement and position in the periodic table			
	Properties and reactions of alkali metals			
	Properties and reactions of halogens			
	Properties and uses of noble gases			
	Properties and uses of transition metals			
3.2 Chemical reactions	Types of chemical reaction			
3.3 Rates of reaction	Reactions at different rates			
	Factors affecting rate of reaction			
	Explaining different rates using particle model			
	Enzymes			
3.4 Reversible reactions	Getting a maximum yield			
	Contact process			
3.5 Energy transfer in reactions	Endothermic and exothermic reactions			
	Bond making and bond breaking			

Chapter

1 Classifying materials

The following topics are covered in this section:

- ***Atomic structure***
- ***Bonding***

What you should know already

Complete the passage, using words from the list. These words should also be used to label the diagram. You can use words more than once.

calcium	**carbon**	**chlorine**	**elements**	**iron**
lead	**liquid**	**magnesium**	**metal**	**non-metal**
oxygen	**periodic table**	**sodium**	**sulphur**	**symbol**

The simplest substances from which all other substances are made up are called: 1.__________.

They are shown in the 2.__________ and can be represented by a chemical 3.__________. This consists of one or two letters. The first letter is always a capital letter.

Complete the table using names from the list:

O	Oxygen	Ca	Calcium	Cu	Copper
C	4.________	Mg	6.________	Fe	8.________
S	5.________	Cl	7.________	Pb	9.________

Elements can be divided into groups in two ways:

- Solid, 10.__________ and gas
- 11.__________ and 12.__________

Elements combine in fixed proportions to form compounds.

The compound sodium chloride is composed of two elements: the metal 13.__________ and the non-metal 14.__________. The compound calcium carbonate contains the metal 15.__________ and two non-metals carbon and 16.__________.

ANSWERS

1. elements; 2. periodic table; 3. symbol; 4. carbon; 5. sulphur; 6. magnesium; 7. chlorine; 8. iron; 9. lead; 10. liquid; 11. and 12. metal and non-metal; 13. sodium; 14. chlorine; 15. calcium; 16. oxygen.

1.1 Atomic structure

LEARNING SUMMARY

After studying this section you should be able to:

- ***recall the particles that make up all atoms and the properties of these particles***
- ***work out the numbers of protons, neutrons and electrons using mass number and atomic number***
- ***describe the structures of atoms of the first 20 elements***
- ***explain why some elements contain different isotopes***
- ***explain the link between reactivity and electron arrangement.***

Particles in an atom

AQA | Edexcel A | Edexcel B | OCR A [A] | OCR A [B] | NICCEA | WJEC

All **elements** are made up from **atoms**.

KEY POINT

An atom is the smallest part of an element that can exist.

About 2000 years ago, Democritus, a Greek philosopher, claimed that all substances were made up of atoms. He had no evidence for this so it was dismissed. Until the early 19th century, scientists believed that atoms were indivisible – like snooker balls. In 1803, Dalton revived the idea of matter being made up of atoms.

It has been found that the atoms of all elements are made up from three basic particles and that the atoms of different elements contain different numbers of these three particles. These particles are:

Particle	Mass	Charge
Proton p	1 u (u is atomic mass unit)	+1
Electron e	Negligible	–1
Neutron n	1 u	Neutral

KEY POINT

Because an atom has no overall charge, the number of protons in any atom is equal to the number of electrons.

In the atom the protons and neutrons are tightly packed together in the **nucleus**. The nucleus is **positively charged**. The electrons move around the nucleus in **energy levels** or **shells**.

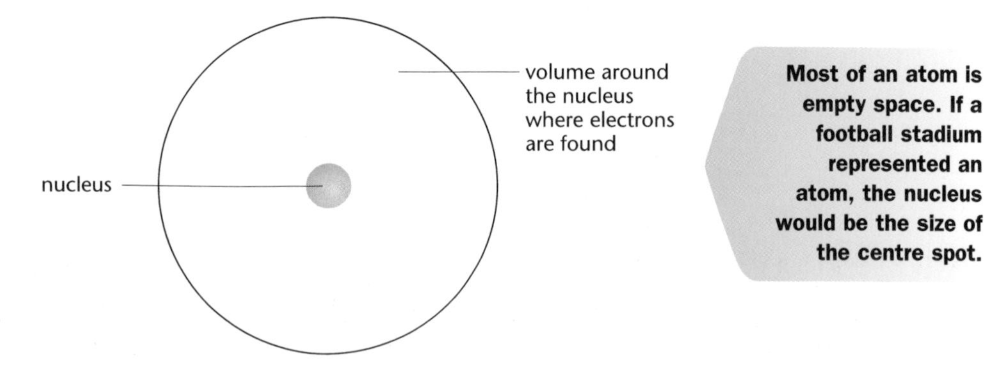

Most of an atom is empty space. If a football stadium represented an atom, the nucleus would be the size of the centre spot.

Fig 1.1 shows a simple representation of an atom.

Atomic number and mass number

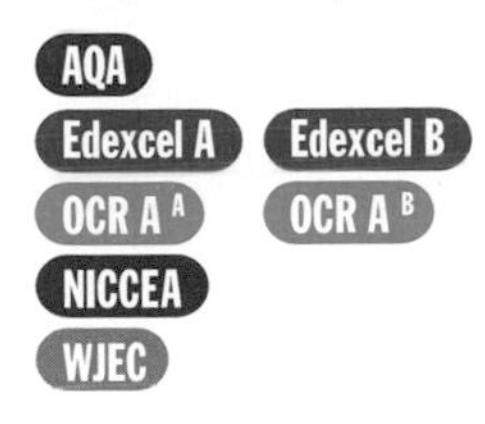

There are two 'vital statistics' for any atom.

- **Atomic number**

The atomic number is the number of **protons** in an atom.

Candidates often get atomic number and mass number confused

- **Mass number**

The mass number is the total number of **protons and neutrons** in an atom.

Atomic number is sometimes called proton number.

We can use these numbers for any atom to work out the number of protons, neutrons and electrons.

E.g. The mass number of carbon-12 is 12, and the atomic number is 6.

Therefore a carbon-12 atom contains 6 protons (i.e. atomic number = 6), 6 electrons and 6 neutrons. This is sometimes written as:

All atoms of the same element have the same atomic number and contain the same number of protons and electrons.

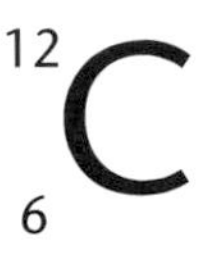

(the atomic number is written under the mass number).

For an atom of sodium-23:

mass number = 23; atomic number = 11

Sodium –23 can be written: $^{23}_{11}Na$

number of protons = 11

number of electrons = 11

number of neutrons = 23 – 11= 12

Isotopes

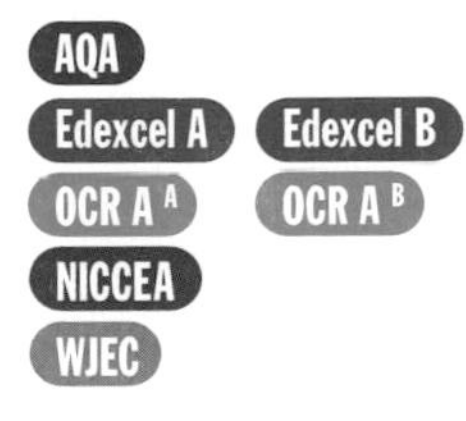

It is possible, with many elements, to get more than one type of atom.

For example, there are three types of oxygen atom:

oxygen-16	8p, 8e, 8n
oxygen-17	8p, 8e, 9n
oxygen-18	8p, 8e, 10n

These three different atoms contain 8 protons and 8 electrons. This determines that all atoms are oxygen atoms.

These different types of atom of the same element are called **isotopes**.

Isotopes are atoms of the same element containing the same number of protons and electrons but different numbers of neutrons.

Isotopes of the same element have the **same chemical properties** but slightly **different physical properties**. There are two isotopes of chlorine — chlorine-35 and chlorine-37. An ordinary sample of chlorine contains approximately 75 per cent chlorine-35 and 25 per cent chlorine-37. This explains the fact that the relative atomic mass of chlorine is approximately 35.5. (The relative atomic mass of an element is the mass of an 'average atom' compared with the mass of a $^{12}_{6}C$ carbon atom.)

Some elements, e.g. fluorine, have only one isotope, but others have different isotopes. Calcium, for example, contains six isotopes.

Arrangement of electrons in an atom

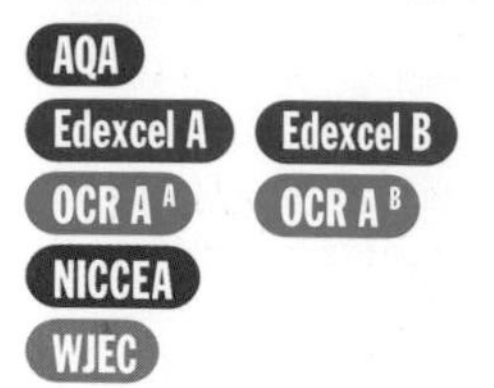

The electrons move rapidly around the nucleus in distinct energy levels. Each energy level is capable of holding only a certain maximum number of electrons. This is represented in a simplified form in **Fig. 1.2**.

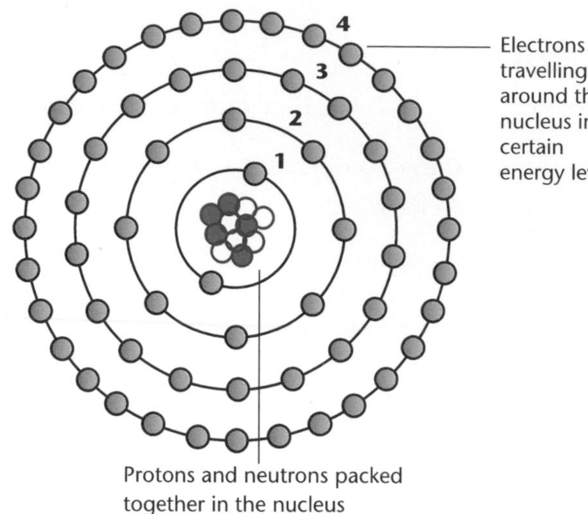

Fig. 1.2 Arrangement of particles in an atom

Beyond element 20 (calcium), the order of filling energy levels is slightly different. This is beyond GCSE.

These energy levels are sometimes called 'shells'.

- The **first energy level** (labelled 1 in **Fig. 1.2**) can hold only **two electrons**. This energy level is filled first.
- The **second energy** level (labelled 2 in **Fig. 1.2**) can hold only **eight electrons**. This energy level is filled after the first energy level and before the third energy level.
- The **third energy** level (labelled 3 in **Fig. 1.2**) can hold a maximum of **18 electrons**. However, when eight electrons are in the third energy level there is a degree of stability and the next two electrons added go into the fourth energy level (labelled 4 in **Fig. 1.2**). Then extra electrons enter the third energy level until it contains the maximum of 18 electrons.
- There are further energy levels, each containing a larger number of electrons than the preceding energy level.

Table 1.1 (see page 13) gives the number of protons, neutrons and electrons in the principal isotopes of the first 20 elements. The arrangement of electrons 2,8,1 denotes 2 electrons in the first energy level, 8 in the second, and 1 in the third. This is sometimes called the **electron arrangement** or **electronic configuration** of an atom.

You ought to be able to draw simple diagrams of atoms of the first 20 elements. Don't forget to show the nucleus and all energy levels.

Atoms are sometimes shown in simple diagrams. The diagram shows a sodium atom.

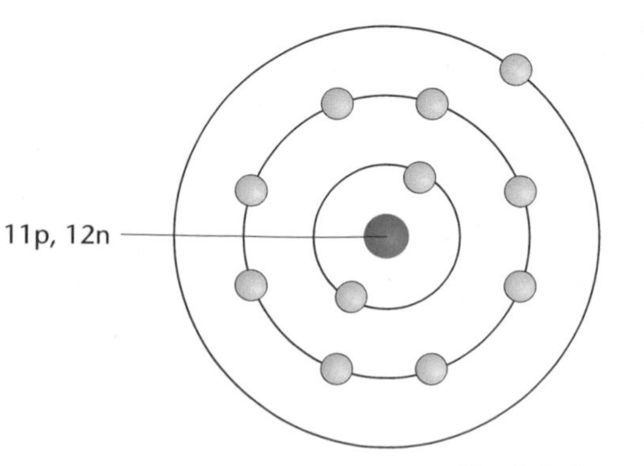

Fig 1.3 A sodium atom

Table 1.1 Numbers of protons, neutrons and electrons in the principal isotopes of the first 20 elements.

Do not try to remember all of the information in the table. You will be able to work it out from the periodic table.

Element	Atomic number	Mass number	Number of			Electron arrangement
			p	n	e	
Hydrogen	1	1	1	0	1	1
Helium	2	4	2	2	2	2
Lithium	3	7	3	4	3	2,1
Beryllium	4	9	4	5	4	2,2
Boron	5	11	5	6	5	2,3
Carbon	6	12	6	6	6	2,4
Nitrogen	7	14	7	7	7	2,5
Oxygen	8	16	8	8	8	2,6
Fluorine	9	19	9	10	9	2,7
Neon	10	20	10	10	10	2,8
Sodium	11	23	11	12	11	2,8,1
Magnesium	12	24	12	12	12	2,8,2
Aluminium	13	27	13	14	13	2,8,3
Silicon	14	28	14	14	14	2,8,4
Phosphorus	15	31	15	16	15	2,8,5
Sulphur	16	32	16	16	16	2,8,6
Chlorine	17	35	17	18	17	2,8,7
Argon	18	40	18	22	18	2,8,8
Potassium	19	39	19	20	19	2.8.8.1
Calcium	20	40	20	20	20	2,8,8,2

Link between reactivity and electron arrangement

AQA
Edexcel A Edexcel B
OCR A [A] OCR A [B]
NICCEA
WJEC

KEY POINT **The reactivity of elements is related to the electron arrangement in their atoms.**

Elements with atoms having **full electron energy** levels are very **unreactive**. These electron arrangements are said to be **stable**. It was believed at one time that they never reacted. These elements include helium, neon and argon.

Elements with atoms containing **one or two electrons** in the outer energy level are very **reactive**. These atoms tend to lose these outer electrons so the atoms finish up with a stable electron arrangement.

Reactive elements have atoms containing nearly empty or nearly full outer energy levels.

Elements with atoms containing **six or seven electrons** in the outer energy level are also **very reactive**. These atoms tend to gain one or more extra electrons so the atoms finish up with, again, a stable electron arrangement.

Elements with atoms containing three, four or five electrons in the outer energy level are usually less reactive.

Hydrogen is interesting. It has an electron arrangement of 1. It is reactive but it can gain 1 electron or lose 1 electron.

Table 1.2 gives some reactive and some unreactive elements. It also gives the arrangement of electrons in atoms of these elements.

Reactive elements		Unreactive elements	
oxygen	2, 6	carbon	2, 4
chlorine	2, 8, 7	silicon	2, 8, 4
fluorine	2, 7	nitrogen	2, 5
- - - - - -	- - - - - -	boron	2, 3
sodium	2, 8, 1		
potassium	2, 8, 8, 1		
calcium	2, 8, 8, 2		

In the reactive elements column, the dotted line separates elements that are reactive because they gain electrons (above the line) from those that are reactive because they lose electrons. From the arrangement of electrons you can make a prediction about whether an element is reactive or unreactive.

PROGRESS CHECK

1. Which particles are always present in equal numbers in an atom?
2. Which particles are in the nucleus of an atom?
3. Iron has an atomic number of 26 and a mass number of 56. What are the numbers of protons, neutrons and electrons in an iron atom?
4. There are three isotopes of hydrogen: Hydrogen-1, Hydrogen-2, Hydrogen-3. How are atoms of these three isotopes different?

Refer back to Table 1.1.
Which of these statements are true and which are false?

5. The elements are arranged in order of increasing atomic number.
6. The number of protons and neutrons is always the same.
7. The number of neutrons is always equal to or greater than the number of protons.
8. The number of neutrons is usually but not always even.
9. Which atom has two filled energy levels?
10. Which atom is shown in the diagram below?

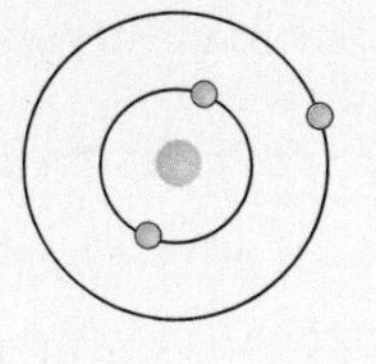

1. Protons and electrons; 2. Protons and neutrons; 3. 26p, 26e, 30n; 4. Different numbers of neutrons: Hydrogen-1 no neutrons, Hydrogen-2 one neutron, Hydrogen-3 two neutrons; 5. True; 6. False; 7. False (look at hydrogen); 8. True; 9. Neon; 10. Lithium.

1.2 Bonding

LEARNING SUMMARY

After studying this section you should be able to:

- ***recall that atoms are joined together by chemical bonds***
- ***understand that ionic bonding takes place when one or more electrons are completely transferred from a metal atom to a non-metal atom***
- ***understand that covalent bonding involves the sharing of pairs of electrons***
- ***describe the differences between giant and molecular structures***
- ***understand that some elements, e.g. carbon, can exist in different forms in the same state. These forms are called allotropes.***

The joining of atoms together is called bonding. An arrangement of particles bonded together is called a structure.

There are several types of bonding found in common chemicals.

Three methods of bonding atoms together are **ionic** bonding, **covalent** bonding and **metallic** bonding.

Ionic (or electrovalent) bonding

KEY POINT **Ionic bonding involves a complete transfer of electrons from one atom to another.**

Two examples are given below:

- **Sodium chloride**

A sodium atom has an electron arrangement of 2,8,1 (i.e. one more electron than the stable electron arrangement of 2,8).

A chlorine atom has an electron arrangement of 2,8,7 (i.e. one electron less than the stable electron arrangement 2,8,8).

Sodium chloride is a compound forced from the reaction of a reactive metal (sodium) and a reactive non-metal (chlorine).

Both the **ions** formed have **stable electron arrangements**. The ions are held together by **strong electrostatic forces**.

Each sodium atom loses one electron to form a sodium ion Na^+. Each chlorine atom gains one electron and forms a chloride ion Cl^-.

It is important to stress in your answers that there is a complete transfer of electrons in ionic bonding. Electrons go from the metal (sodium) to the non-metal (chlorine). A frequent mistake is to use terms such as atoms swapping electrons. This is wrong!

This process is summarised in **Fig. 1.4.**

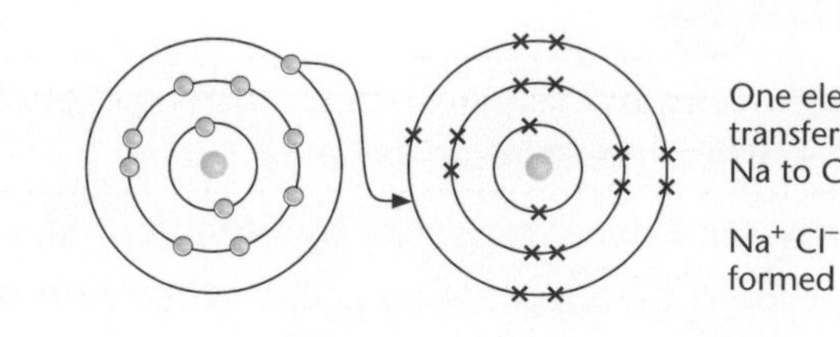

Fig. 1.4 Ionic bonding in sodium chloride

It is incorrect to speak of a 'sodium chloride **molecule**'. This would assume that one sodium ion joins with one chloride ion.

KEY POINT **A sodium chloride crystal consists of a regular arrangement of equal numbers of sodium and chloride ions. This is called a lattice.**

Fig 1.5 A sodium chloride lattice

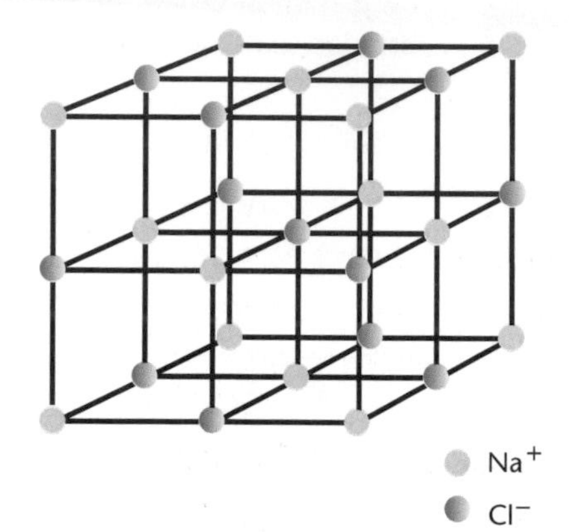

You are only required to know about ionic lattices if you are doing Higher tier.

In ionic bonding, one element is a metal and one is a non-metal. Metal atoms lose electrons and non-metal atoms gain them.

- **Magnesium oxide**

Electron arrangement in magnesium atom 2,8,2

Electron arrangement in oxygen atom 2,6

KEY POINT **Two electrons are lost by each magnesium atom to form Mg^{2+} ions. Two electrons are gained by each oxygen atom to form O^{2-} ions.**

This is summarised in **Fig. 1.6.**

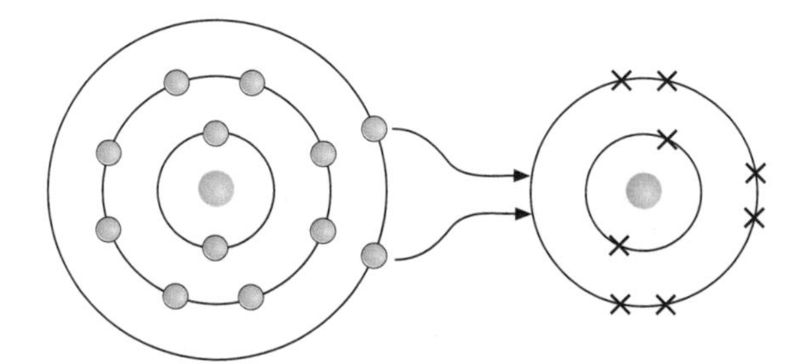

Fig. 1.6 Ionic bonding in magnesium oxide

Loss of one or two electrons by a metal during ionic bonding is common, e.g. NaCl or MgO.

If three electrons are lost by a metal, the resulting compound shows some covalent character, e.g. $AlCl_3$.

Covalent bonding

AQA
Edexcel A Edexcel B
OCR A[A] OCR A[B]
NICCEA
WJEC

Covalent bonding involves the sharing of electrons, rather than complete transfer.

Two examples are given below:

- **Chlorine molecule (Cl_2)**

A chlorine atom has an electron arrangement of 2,8,7. When two chlorine atoms bond together they form a chlorine **molecule**. If one electron was transferred from one chlorine atom to the other, only one atom could achieve a stable electron arrangement.

A similar covalent bond exists in a hydrogen molecule.

Instead, one electron from each atom is donated to form a **pair of electrons** which is shared between both atoms, holding them together. This is called a **single covalent** bond. **Fig. 1.7** shows a simple representation of a chlorine molecule using a dot and cross diagram.

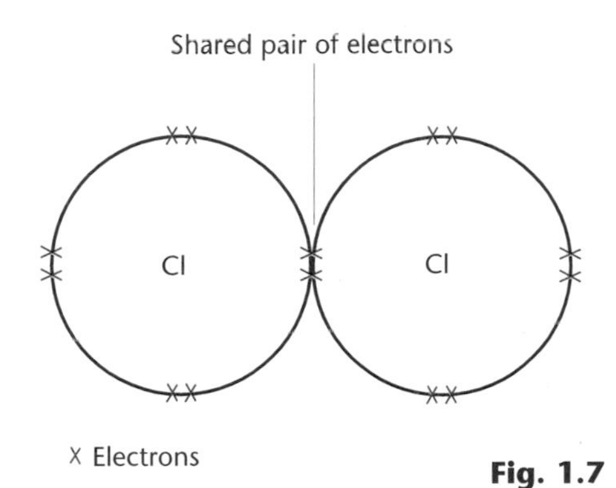

Fig. 1.7

This is often shown as Cl—Cl.

- **Oxygen molecule (O_2)**

An oxygen atom has an electron arrangement of 2,6. In this case each oxygen atom donates two electrons and the **four electrons (two pairs)** are **shared** between both atoms. This is called a **double covalent bond**. **Fig. 1.8** shows a simplified representation of an oxygen molecule.

When drawing dot and cross diagrams do not draw them too small. They can be difficult for the examiner to interpret. Remember both dots and crosses represent electrons.

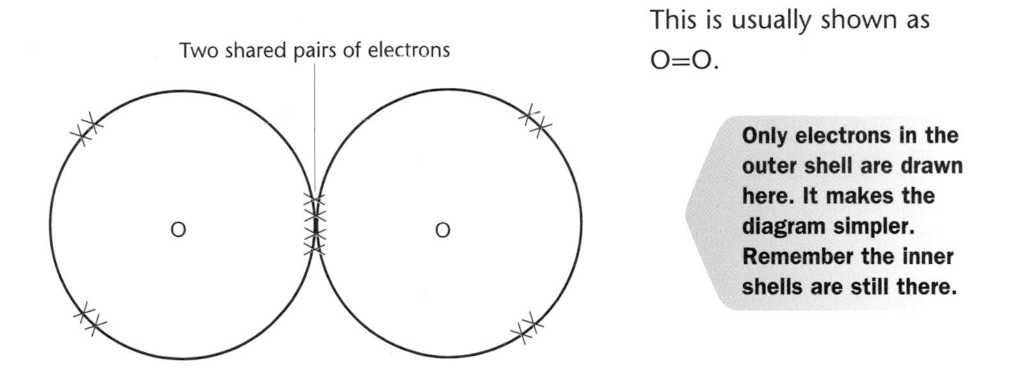

This is usually shown as O=O.

Only electrons in the outer shell are drawn here. It makes the diagram simpler. Remember the inner shells are still there.

Fig. 1.8 below shows other examples of molecules containing covalent bonding.

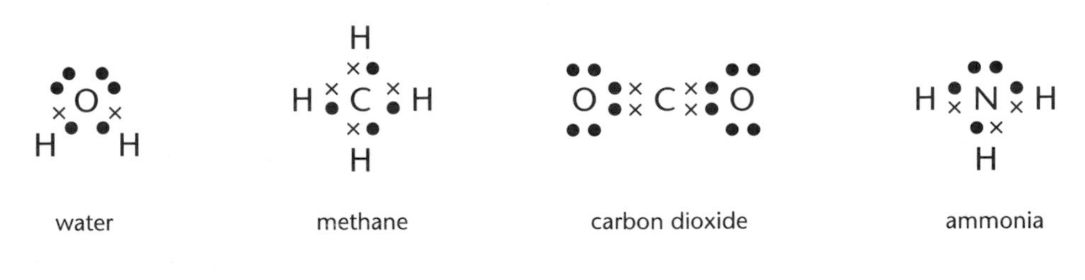

Fig. 1.8

Metallic bonding

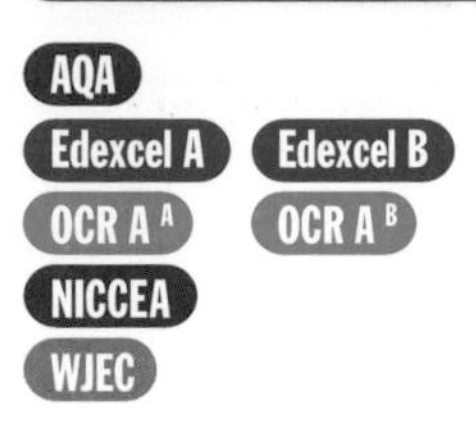

Metallic bonding is found only in metals.

> **KEY POINT** **A metal consists of a close-packed regular arrangement of positive ions, which are surrounded by a 'sea' of electrons that bind the ions together.**

Fig. 1.9 shows the arrangement of ions in a single layer.

The sea of electrons can move throughout the structure. This explains the high electrical conductivity of solid metals. Metals are crystalline. This is due to the regular arrangement of particles in the structure.

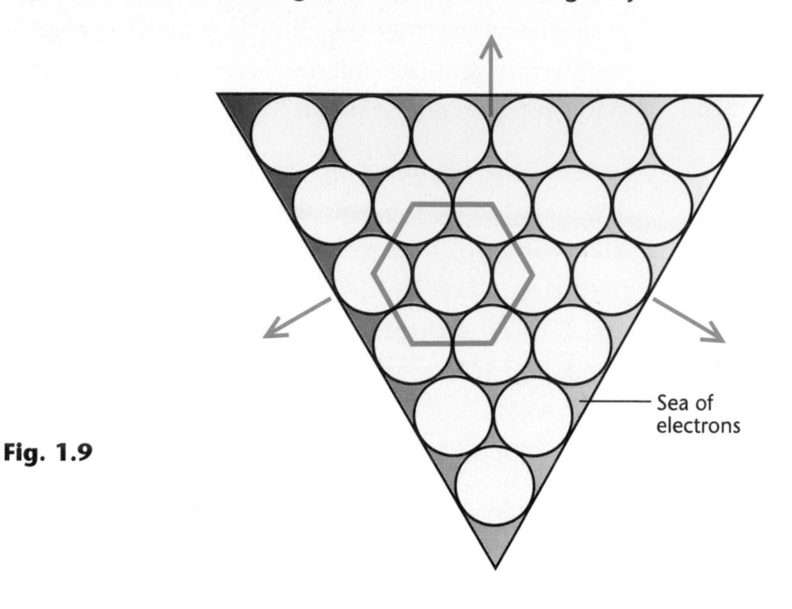

Fig. 1.9

There are two alternative ways of stacking these layers. The arrows in **Fig. 1.9** indicate that the layer shown continues in all directions. Around any one ion in a layer there are six ions arranged hexagonally.

Giant and molecular structures

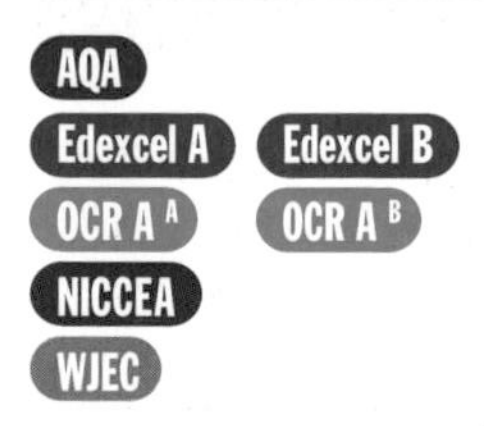

Silicon dioxide, SiO_2, and carbon dioxide, CO_2, both contain covalent bonding to join the atoms together. However, **silicon dioxide is a solid** and **carbon dioxide is a gas**.

In carbon dioxide, each carbon atom is joined with two oxygen atoms to form a **molecule**.

Giant and molecular structures are Higher tier only.

The molecules are not held together.

In silicon dioxide there is, in effect, one large molecule. Each silicon is bonded to four oxygen atoms and each oxygen is bonded to two silicon atoms. The resulting structure is called a **giant structure**.

In a molecular structure there may be strong forces within each molecule but the forces between the molecules are very weak.

Fig. 1.10 shows simple representations of molecular and giant structures.

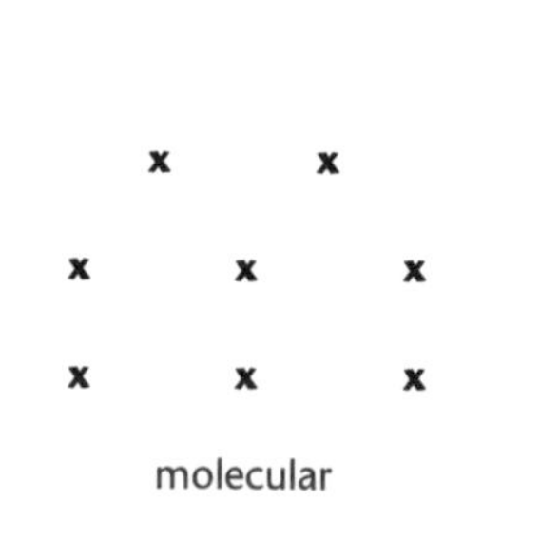

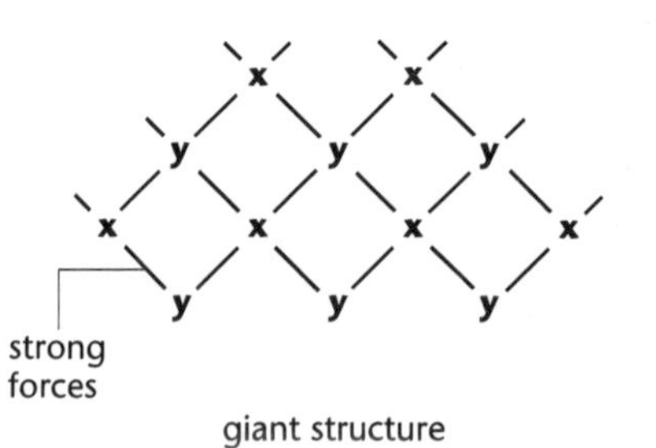

Fig. 1.10

Bonding, structure and properties

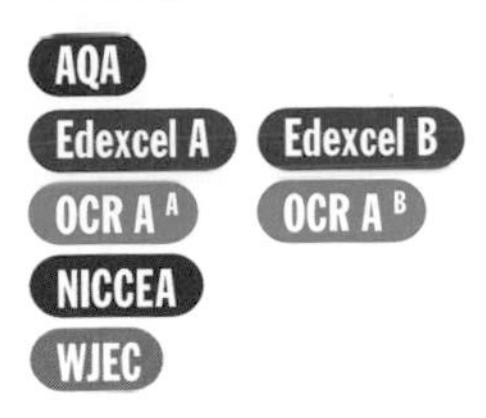

Table 1.3 below summarises how bonding and structure alter properties of substances.

Bonding	Structure	Properties
Ionic	**Giant structure**, e.g. sodium chloride, magnesium oxide	High melting and boiling point, usually soluble in water but insoluble in organic solvents. Conduct electricity when molten or dissolved in water **(electrolytes)**
Covalent	**Molecular**, e.g. chlorine, iodine, methane	Usually gases or low boiling point liquids.Some (iodine and sulphur)are low melting point solids. Usually insoluble in water but soluble in organic solvents. Do not conduct electricity
	Macromolecules (large molecules) e.g. poly(ethene), starch	Solids. Usually insoluble in water but more soluble in organic solvents. Do not conduct electricity
	Giant structure, e.g. silicon dioxide	Solids. High melting points. Insoluble in water and organic solvents. Do not conduct electricity
Metallic	**Giant structure**, e.g. copper	Solids. High density (ions closely packed). Good electrical conductors (free electrons)

Solid sodium chloride does not conduct electricity. The ions are not free to move. In molten sodium chloride and sodium chloride solution the ions are free to move and they conduct electricity.

Allotropy

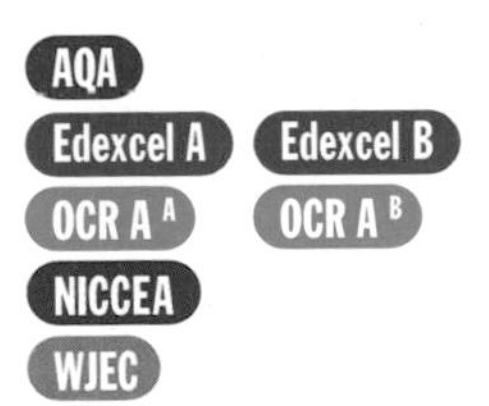

KEY POINT **Allotropy is the existence of two or more forms of an element in the same physical state.**

Solid lead and molten lead are not allotropes because they are not in the same physical state – one is solid and the other is liquid.

These different forms are called **allotropes**. Allotropy is caused by the possibility of more than one arrangement of atoms. For example, carbon can exist in allotropic forms including **diamond** and **graphite**. Sulphur can exist in two allotropes — α-**sulphur** and β-**sulphur**.

Oxygen, O_2, and ozone, O_3, are two gaseous forms of oxygen. They are allotropes. You have probably heard of the ozone layer.

Allotropy of carbon

KEY POINT **The two most commonly mentioned allotropes of carbon are diamond and graphite.**

Another allotrope of carbon is **fullerene** which is a crystalline form of carbon made of clusters of carbon atoms.

1 Diamond

In the diamond structure, each carbon atom is strongly bonded (covalent bonding) to four other carbon atoms tetrahedrally

A large **giant structure** (three-dimensional) is built up. All bonds between carbon atoms are the same length (0.154 nm). It is the strength and uniformity of the bonding which make diamond very hard, non-volatile and resistant to chemical attack. **Fig. 1.11** shows the arrangement of particles in diamond.

Other elements show allotropy including sulphur, phosphorus and tin.

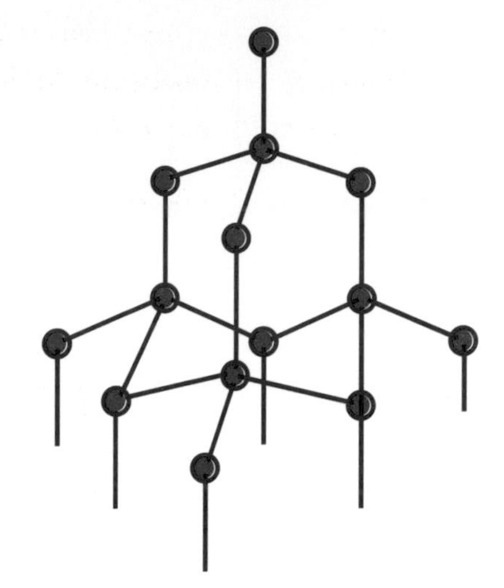

Fig. 1.11

2 Graphite

Graphite has a **layer structure**. In each layer the carbon atoms are bound covalently. The bonds **within the layers** are very **strong**.

The bonds between the layers of graphite are very weak, which enables layers to slide over one another.

This makes the graphite soft and flaky.

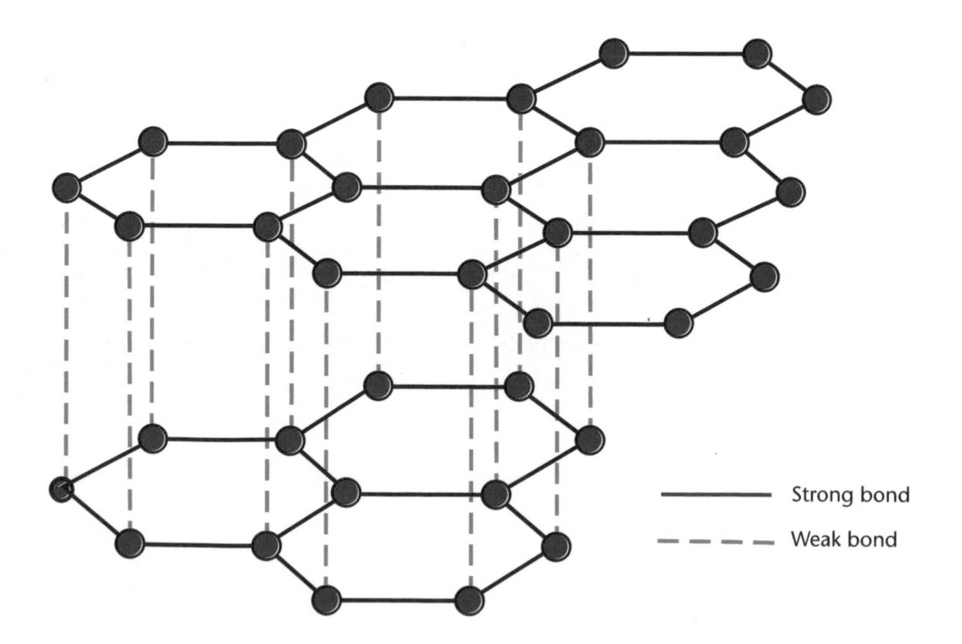

Fig. 1.12 Structure of graphite

Table 1.4 compares the properties of diamond and graphite.

Property	Diamond	Graphite
Appearance	Transparent, colourless crystals	Black, opaque, shiny solid
Density (g/cm^3)	3.5	2.2
Hardness	Very hard	Very soft
Electrical conductivity	Non-conductor	Good electrical conductor

Many other important discoveries have been made by accident, e.g. poly(ethene), xenon tetrafluoride.

A chance discovery in 1985 led to the identification of a new allotrope of carbon. In fact, a new family of closed carbon clusters has been identified and called **fullerenes**. Two fullerenes, C_{60} and C_{70}, can be prepared by electrically evaporating carbon electrodes in helium gas at low pressure. They dissolve in benzene to produce a red solution.

Fig. 1.13 shows a fullerene molecule.

Fullerenes are good lubricants as molecules can easily slide over each other.

The process for making fullerenes has to be carried out in atmosphere of helium. In air the carbon would burn.

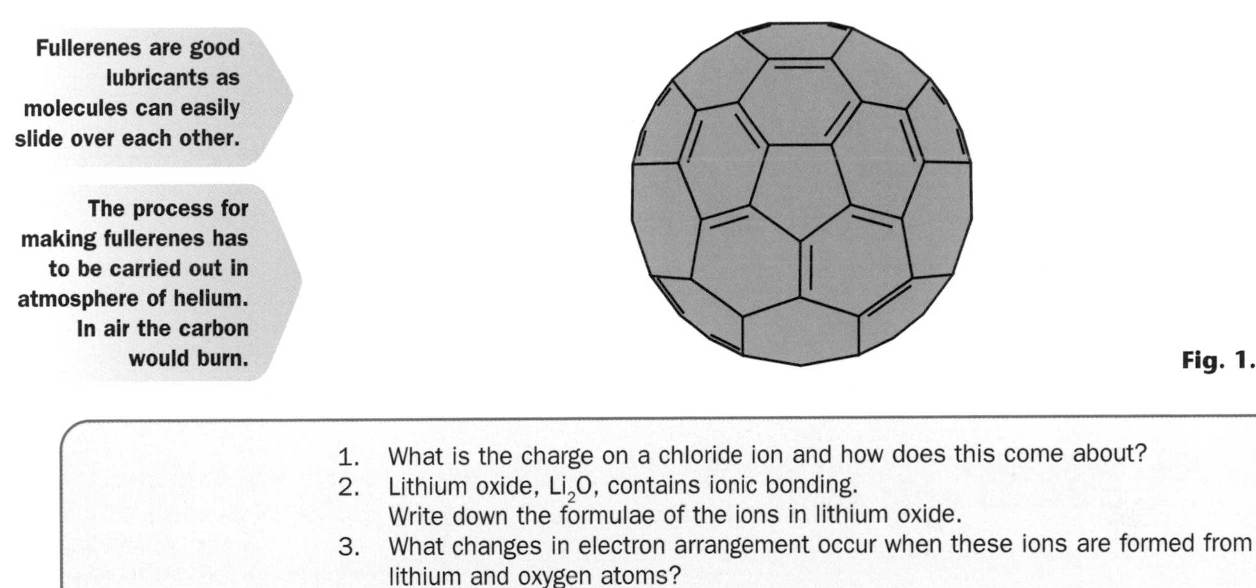

Fig. 1.13

PROGRESS CHECK

1. What is the charge on a chloride ion and how does this come about?
2. Lithium oxide, Li_2O, contains ionic bonding.
   Write down the formulae of the ions in lithium oxide.
3. What changes in electron arrangement occur when these ions are formed from lithium and oxygen atoms?
4. What type of forces hold these ions together in the solid?
5. Some atoms complete their shells by sharing electrons. What type of bonding is this?
   Use this list to answer questions 6–9
   diamond **magnesium oxide** **methane** **silicon dioxide**
6. Which substance in the list is an example of an element with a giant structure of atoms?
7. Which substance in the list is an example of a giant structure of ions?
8. Which substance in the list is an example of a compound with a giant structure of atoms?
9. Which substance in the list is an example of a molecular structure?
10. Why are metals good conductors of electricity?

1. One negative charge – gains one electron; 2. Li^+ and O^{2-}; 3. Lithium atom loses one electron, oxygen atom gains two electrons; 4. Electrostatic; 5. Covalent; 6. Diamond; 7. Magnesium oxide; 8. Silicon dioxide; 9. Methane; 10. Free electrons move through the metal.

Sample GCSE questions

1. Europium is an element discovered in 1901 by E.A. Demarcay in France.

It is a silvery-white metal.
It was given the symbol Eu.
Its atoms have an electron arrangement 2, 8, 18, 25, 8, 2.

(a) What is the atomic number of europium? **[1]**

63 ✓

Just count up the number of the electrons. This is the same as the number of protons and is equal to the atomic number.

(b) There are two isotopes of europium. They are europium-151 and europium-153. In a sample of europium there is 50% of each isotope.

(i) How many neutrons are there in each isotope? **[2]**

Europium-151 88 neutrons ✓
Europium-153 90 neutrons ✓

You can work out the number of neutrons by subtracting the atomic number from the mass number.

(ii) What is the relative atomic mass of europium? Explain your answer. **[2]**

152 ✓
Half way between 151 and 153 because the isotopes are present in equal amounts ✓.

(iii) How many particles are present in the nucleus of a europium-151 atom? **[1]**

151 – protons plus neutrons ✓

The particles in the nucleus are sometimes called nucleons.

2. The table shows the number of protons and electrons in sodium and fluorine atoms.

Atom	Number of protons	Number of electrons
Sodium	11	11
Fluorine	9	9

(a) Draw diagrams to show the arrangement of electrons in a sodium atom and in a fluorine atom. **[2]**

sodium ✓ *flourine ✓*

The periodic table could help you here.

Sample GCSE questions

(b) **(i)** Draw a diagram to show outer electrons in a fluorine molecule, F_2. **[2]**

✓✓

> *Make sure you understand the differences between ionic and covalent bonding.*

(ii) What type of bonding is present in a fluorine molecule? **[1]**

Covalent bonding ✓

(c) When sodium and fluorine combine, electron transfer takes place and ions are formed.

(i) What electron transfer takes place? **[2]**

Sodium atom loses an electron ✓
Fluorine atom gains an electron ✓

(II) Write down one similarity and one difference between sodium and fluoride ions. **[2]**

Similarity: ✓ Same number of electrons (or same electron arrangement) ✓
Difference: ✓ Different number of neutrons ✓

> *Other differences include different numbers of protons, different charges or different atomic radii.*

(d) Sodium fluoride has a **giant structure**.

(i) What is a giant structure? **[2]**

All of the particles joined together to form a single network or structure. ✓ In this case there is a giant structure of ions rather than atoms ✓.

(ii) Suggest two properties of sodium fluoride **[2]**

High melting point ✓
Conducts electricity when molten or in aqueous solution ✓

> *Dissolves in water would be another correct answer.*

Exam practice questions

1. Hydrogen and chlorine react together to form hydrogen chloride.

$$H_2 + Cl_2 \rightarrow 2HCl$$

(a) Dry hydrogen chloride gas contains hydrogen chloride molecules.

(i) Draw a dot and cross diagram of a hydrogen chloride molecule. **[2]**

(ii) What type of bonding is present in dry hydrogen chloride molecules? **[1]**

(b) Hydrogen chloride dissolves in water to form a solution which conducts electricity.

Explain the changes in bonding which occur when hydrogen chloride dissolves in water. **[3]**

2. The table below gives information about four substances labelled A–D.

Substance	Melting point in °C	Boiling point in °C	Electrical conductivity when solid	Electrical conductivity when molten
A	801	1470	poor	good
B	850	1487	good	good
C	–218	–183	poor	poor
D	2900	very high	poor	poor

What does the data in the table show about the structures of these substances?

Explain your reasoning. **[8+1]**

{This question is marked out of eight marks for the correct science in your answer. In addition one mark is allocated for the examiner to use to reward some aspect of Quality of Written Communication (QWC). In this case the mark will be awarded for an answer written in proper sentences with a capital letter and a full stop.}

Chapter 2

Changing materials

The following topics are covered in this section:

- *Chemicals from organic sources*
- *Useful products from metal ores and rocks*
- *Useful products from air*
- *Quantitative chemistry*
- *Earth cycles*

What you should know already

Complete the passage, using words from the list. These words should also be used to label the diagram. You can use words more than once.

acid rain	**chemical**	**combustion**	**fossil fuels**	**igneous rocks**
increases	**magma**	**mass**	**metamorphic rocks**	**physical**
reversed	**saturated**	**sedimentary rocks**	**solute**	**solvent**

Melting, freezing, grinding and dissolving are 1.__________ changes that can easily be 2.__________. When such a change occurs, there is no change in 3.__________.
Changes where new substances are produced by chemical reactions are called 4.__________ changes. These changes are not easily reversed. An example of this type of change is burning (or 5.__________).
A substance that dissolves is called a 6.__________ and the liquid in which it dissolves is called the 7.__________. The mixture is called a solution. A solution that contains the maximum amount of dissolved solid is called a 8.__________ solution. The solubility of a solute usually 9.__________ with rise in temperature.

There are three types of rock – sedimentary, igneous and metamorphic.

Use words in the list to label the diagram of the rock cycle.

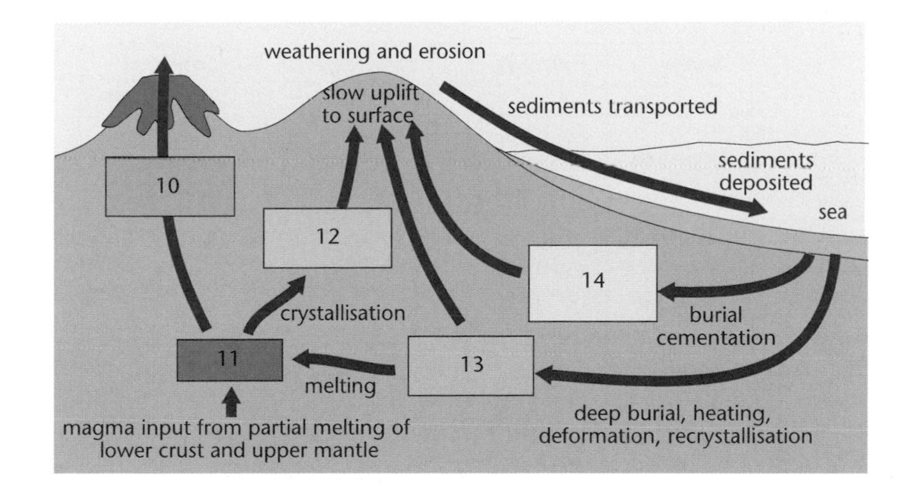

Fuels such as coal oil and gas are called 15.__________ fuels. When they burn they form products such as carbon dioxide and also gases such as sulphur dioxide which can produce 16.__________.

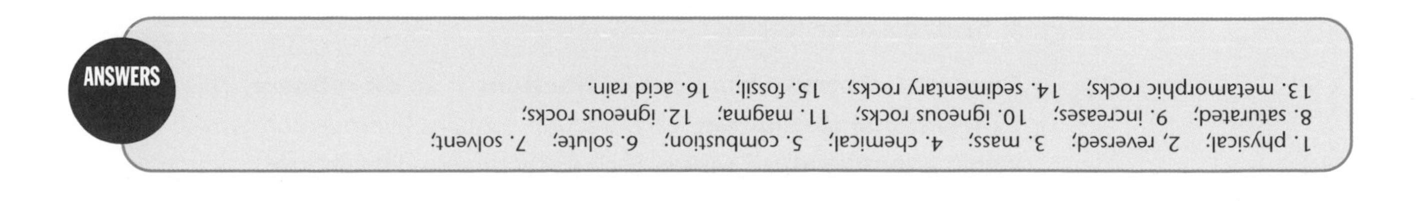

ANSWERS

1. physical; 2. reversed; 3. mass; 4. chemical; 5. combustion; 6. solute; 7. solvent; 8. saturated; 9. increases; 10. igneous rocks; 11. magma; 12. igneous rocks; 13. metamorphic rocks; 14. sedimentary rocks; 15. fossil; 16. acid rain.

2.1 Chemicals from organic sources

LEARNING SUMMARY

After studying this section you should be able to:

- ***understand how crude oil can be split up into saleable products by fractional distillation***
- ***understand there are different families of hydrocarbons including alkanes and alkenes***
- ***explain that different products are formed when hydrocarbons burn in different amounts of oxygen***
- ***understand that large chain hydrocarbons can be broken into simpler molecules by cracking***
- ***explain how small alkenes can be linked together to form addition polymers***
- ***evaluate the benefits of addition polymers for a range of uses.***

Refining crude oil

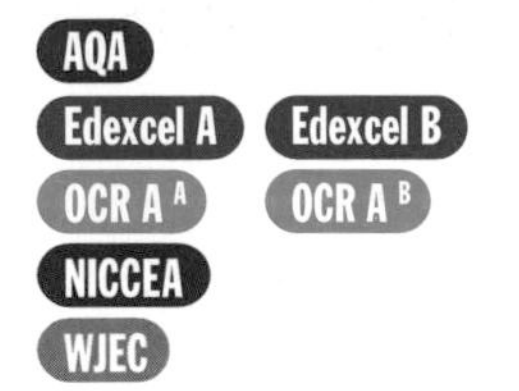

Crude oil is a **mixture** of **hydrocarbons**. Hydrocarbons are **compounds** of **carbon** and **hydrogen only**.

Most of the hydrocarbons belong to a family called **alkanes**.

Table 2.1 contains information about the first six members of the alkane family.

Name	Formula	Structure	State at room temp	Melting point	Boiling point
Methane	CH_4	H | H–C–H | H	Gas	Increases down the family ↓	Increases down the family ↓
Ethane	C_2H_6	H H | | H–C–C–H | | H H	Gas		
Propane	C_3H_8	H H H | | | H–C–C–C–H | | | H H H	Gas		
Butane	C_4H_{10}	H H H H | | | | H–C–C–C–C–H | | | | H H H H	Gas		
Pentane	C_5H_{12}	H H H H H | | | | | H–C–C–C–C–C–H | | | | | H H H H H	Liquid		

Remember the names of alkanes end in -ane. The prefix tells you the number of carbon atoms. Pentane contains 5 carbon atom.

Alkanes

- are all **saturated hydrocarbons** (contains only single carbon–carbon bonds)
- all fit a formula C_nH_{2n+2}
- burn in air or oxygen
- have few other reactions.

Candidates often confuse saturated here with saturated when referred to solutions.

Crude oil is separated into separate **fractions** in an **oil refinery**. This is done by **fractional distillation**. Each fraction contains hydrocarbons which boil within a **temperature range**. Each fraction has a different use.

Crude oil is sometimes called petroleum.

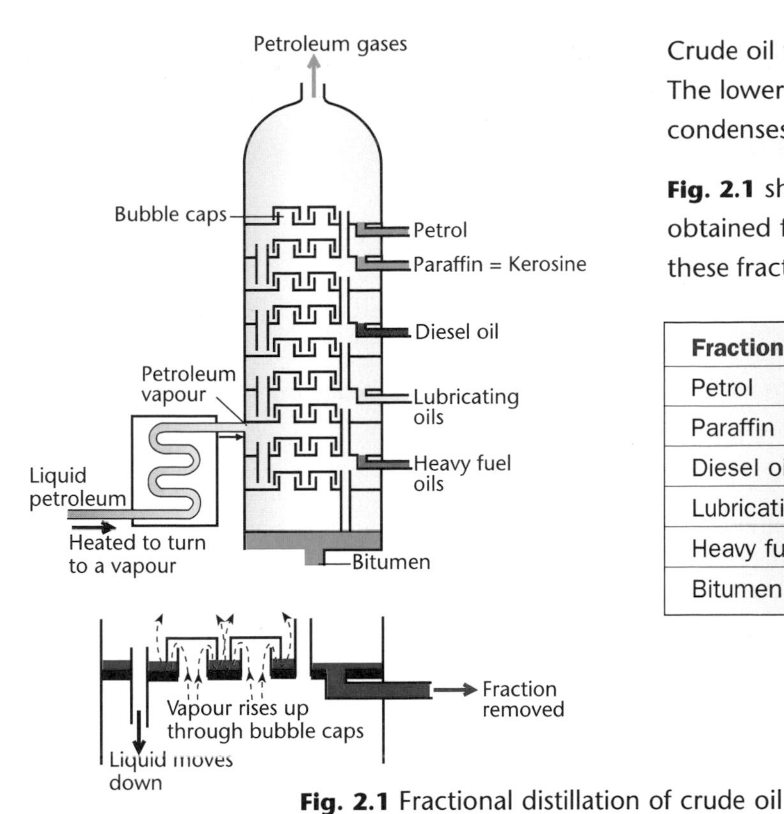

Fig. 2.1 Fractional distillation of crude oil

Crude oil vapour enters a tall column and cools. The lower the boiling point the higher the vapour condenses in the column.

Fig. 2.1 shows how different fractions can be obtained from crude oil, and the different uses of these fractions are shown in the table below.

Fraction	Use
Petrol	Fuel for cars
Paraffin	Fuel for aircraft
Diesel oil	Fuel for cars, trains
Lubricating oil	For motor engines
Heavy fuel oils	For heating
Bitumen	Road tar

Burning alkanes

When alkanes are burned in **excess air** or oxygen, **carbon dioxide** and **water** are produced.

e.g. methane + oxygen → carbon dioxide + water

$CH_4 \quad + 2O_2 \quad \rightarrow CO_2 \quad + 2H_2O$

Alkanes burn in a **limited supply of air** to produce **carbon monoxide** and **water vapour**. Carbon monoxide is very **poisonous**.

e.g. methane + oxygen → carbon monoxide + water

$2CH_4 \quad + 3O_2 \quad \rightarrow 2CO \quad + 4H_2O$

Carbon monoxide has no smell. Every year in the UK up to 50 people die of carbon monoxide poisoning. Often these deaths are caused by gas appliances with inadequate ventilation.

Making addition polymers

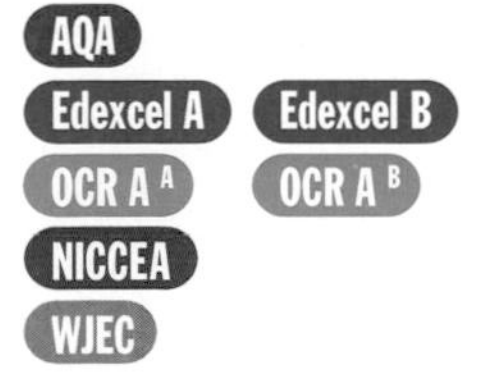

Higher boiling point fractions are more **difficult to sell** as there is less demand for them.

The petrochemical industry breaks up these long chains to produce short molecules. This decomposition reaction is called **cracking**.

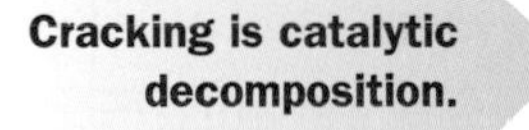

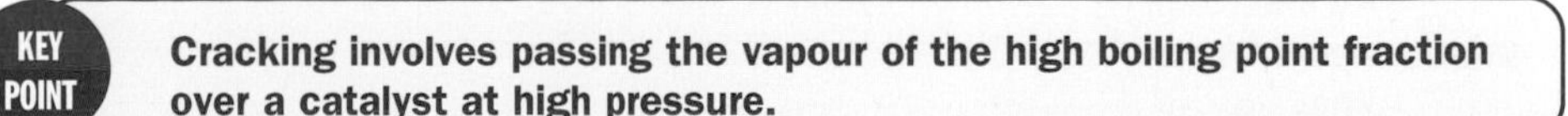

Compounds such as **ethene** are produced.

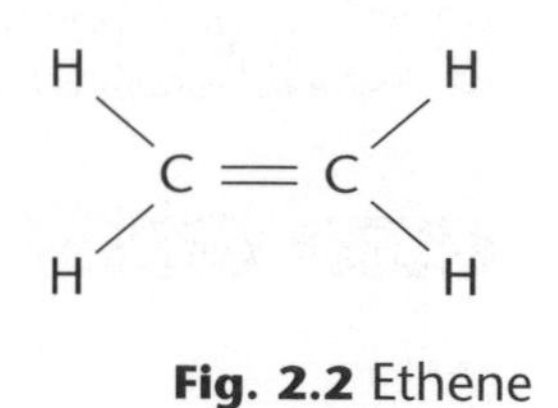

Fig. 2.2 Ethene

Ethene

- belongs to a family of **alkenes**, C_nH_{2n}.
- is an **unsaturated hydrocarbon** with a formula C_2H_4.
- is a gas at room temperature.
- molecules contain a **carbon–carbon double bond**.

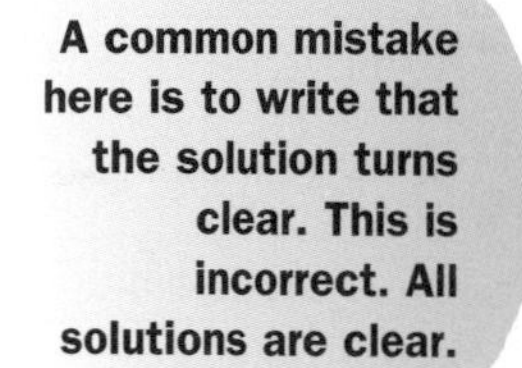

There is a simple test to distinguish ethane and ethene. If ethene gas is bubbled through a solution of **bromine** the solution changes from **red–brown** to **colourless**.

ethene + bromine → 1,2-dibromoethane

$C_2H_4 + Br_2 \rightarrow C_2H_4Br_2$

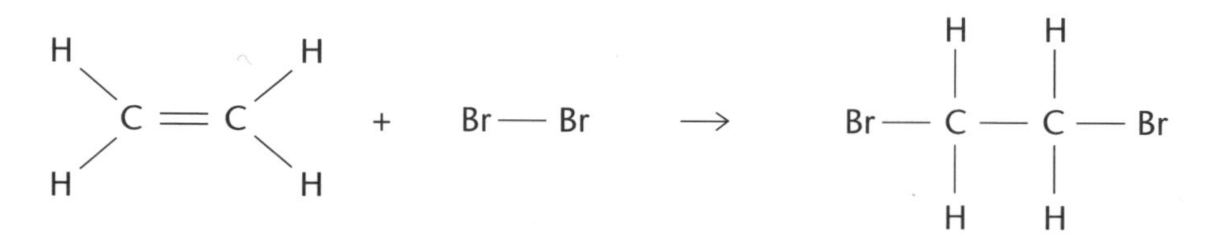

Fig. 2.3 Addition of bromine to ethene

In an addition reaction two molecules combine to form a single product.

This is an example of an **addition reaction**. Two reactants react to form a single product and the double bond in ethene becomes a single bond.

There is no colour change when ethane is added to a solution of bromine.

Small ethene molecules, produced by cracking, are joined together by a process called addition polymerisation.

Ethene is called the **monomer** and **poly(ethene)** is called the **addition polymer**. In order to produce this polymer, the ethene vapour is passed over a heated catalyst. A series of addition reactions occur.

Notice that the monomer contains a double bond and this becomes a single bond when the molecules join together. The chains can have thousands of units added together. The properties of a sample of polymer depend upon chain length.

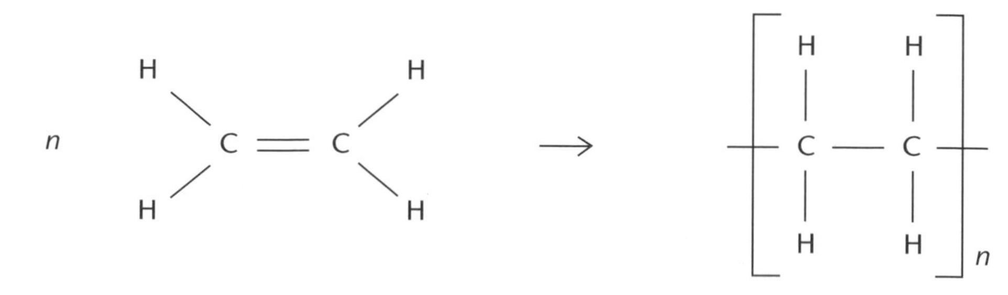

Fig. 2.4 Polymerisation of ethene

Uses of addition polymers

Addition polymers such as poly(ethene) and poly(vinyl chloride) have many uses. They have replaced traditional materials such as metals, paper, cardboard and rubber.

Common uses include:

poly(ethene) – wrappings for food, storage containers, milk crates

poly(vinyl chloride) – wellington boots, insulation for electrical wiring

Table 2.2 compares some of the advantages and disadvantages of polymers.

Advantages of polymers	Disadvantages of polymers
Do not absorb water	Do not rot away and can cause litter problems
Can be moulded into shape	Not easy to recycle as there are many types
Can be coloured	Burn to form poisonous fumes
Low density	
Strong	

PROGRESS CHECK

1. Which one of the following compounds is not a hydrocarbon?
   C_2H_4 C_2H_6O C_6H_6 C_4H_{10}
2. Which one of the hydrocarbons in the list is not an alkane?
   C_6H_{12} C_7H_{16} $C_{10}H_{22}$ $C_{40}H_{82}$
3. Write down a use for each of the following alkanes:
   (a) methane; (b) propane; (c) octane
4. LPG is used as a fuel in cars as an alternative to petrol. How is it produced in the refining process?
5. A colourless gas **X** has a formula C_3H_6. It decolorises bromine.
   X could be **A.** ethane **B.** ethene **C.** propane **D.** propene
6. Poly(vinyl chloride) is made from a monomer called vinyl chloride:

   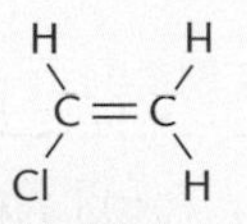

   Which is the correct chemical name for vinyl chloride?
   chloroethane chloroethene ethene
7. Draw the structure of poly(vinyl chloride).
8. Three fractions from the crude oil refinery are: kerosene, petrol and bitumen. Put these three fractions in the correct order working down from the top of the column.
9. Suggest uses for each of the fractions in 8.
10. Decane, $C_{10}H_{22}$ can be cracked into a mixture of ethene and ethane. Write a balanced equation for this reaction.

1. C_2H_6O; 2. C_6H_{12}; 3. (a) Household gas supply (natural gas) (b) Camping gas (c) Petrol;
4. Gases leaving the top of column; 5. D; 6. Chloroethene ;
7. $\left[-\overset{H}{\underset{H}{C}}-\overset{H}{\underset{Cl}{C}}- \right]_n$
8. Petrol, kerosene, bitumen; 9. Fuel in cars; fuel in aeroplanes; road tar
10. $C_{10}H_{22} \rightarrow 4C_2H_4 + C_2H_6$

2.2 Useful products from metal ores and rocks

LEARNING SUMMARY

After studying this section you should be able to:

- ***recall some materials made from rocks and the uses of these materials***
- ***understand how salt and limestone can be used to make useful materials***
- ***understand how metals can be extracted from metal ores***
- ***understand the chemical principles involved in the extraction of aluminium and iron and the purification of copper by electrolysis.***

Products made from rocks

AQA Edexcel A Edexcel B OCR A[A] OCR A[B] NICCEA WJEC

The rocks of the Earth are the source of a wide range of materials.

Rocks as building materials

Rocks such as **limestone**, **sandstone** and **slate** are used as **building materials**. Because transport costs are very high, rocks are often used as building materials close to where they are dug out of the ground **(quarried)**.

Building materials made from rocks

Because natural rocks are expensive, new materials have been developed to replace them. **Table 2.3** summarises how some of these materials are made.

Table 2.3 Building materials made from rocks

Material	How it is made	More information
Bricks	By baking clay to a high temperature	Hard and brittle – regular shape
Mortar	Mixture of **calcium hydroxide**, **sand** and **water** made into a thick paste	It sets by losing water and by absorbing carbon dioxide from the air. Long crystals of calcium carbonate are formed
Cement	Heating **limestone** with **clay** (containing aluminium and silicates)	It consists of a complex mixture of calcium and aluminium silicates. When it is mixed with water, a chemical reaction occurs producing calcium hydroxide, and this sets in a similar way to mortar.
Concrete	Made by mixing cement with sand and small stones	Used to make many objects such as drain covers that were previously made from cast iron. Concrete can be strengthened by steel reinforcing rods
Glass	Mixing **limestone**, **sand** (silicon dioxide) and **sodium carbonate** together and melting the mixture	The resulting mixture of calcium and sodium silicates cools to produce glass. Coloured glass is due to transition metal oxides present in the mixture

Chemicals from rocks

Many chemicals are made from rocks.

Limestone is used to make quicklime (calcium oxide) and slaked lime (calcium hydroxide).

Limestone is the mineral extracted from the Earth in the largest amounts. Often, it is found in beautiful areas and its mining can damage the environment.

These processes are summarised in **Fig. 2.5**.

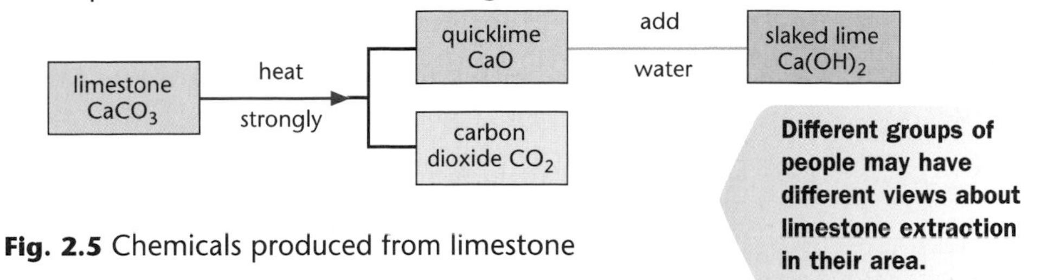

Fig. 2.5 Chemicals produced from limestone

Different groups of people may have different views about limestone extraction in their area.

Rock salt is used as the raw material for producing a wide range of chemicals including sodium hydroxide, chlorine, hydrogen, sodium and household bleach.

Salt is mined by solution mining. Water is pumped underground. The salt dissolves and salt solution (brine) is pumped to the surface.

These processes are summarised in **Fig. 2.6**.

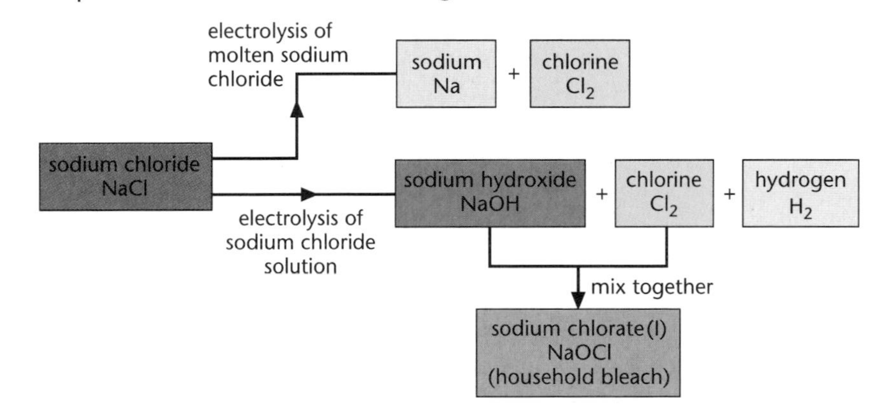

Fig. 2.6 Chemicals produced from rock salt

Metals from rocks

Most metals are found in the earth as deposits of **ore**.

An ore is a rock that contains enough of a metal compound for it to be worth extracting the metal

Metals are in order of reactivity. Notice most reactive metals are present as chlorides,carbonates or oxides and less reactive metals as sulphides.

Table 2.4 Common metal ores and their chief compound

Metal	Name of ore	Compound of metal present
Sodium	Rock salt (halite)	Sodium chloride
Magnesium	Magnesite	Magnesium chloride
Aluminium	Bauxite	Aluminium oxide
Iron	Haematite	Iron(III) oxide
Zinc	Zinc blende	Zinc sulphide
Mercury	Cinnabar	Mercury(II) sulphide

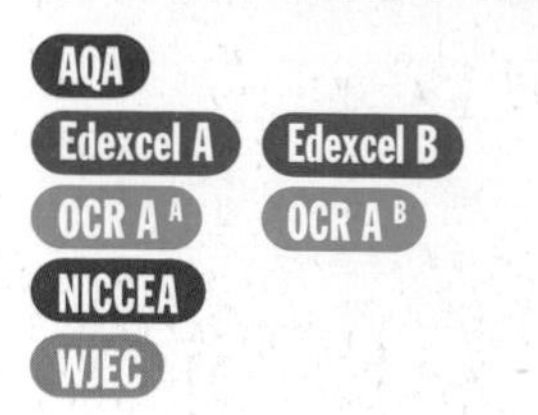

You will certainly need to know the names of the ores of aluminium and iron.

Some ores contain only small amounts of metal compounds. The metal compound in these ores may be concentrated by **froth flotation** before the metal is extracted. The ore is added to a detergent bath and the mixture agitated. By careful control of the conditions, it is possible to get the metal compound to float while the impurities sink to the bottom.

Extracting metals from ores

AQA Edexcel A Edexcel B OCR A[A] OCR A[B] NICCEA WJEC

The method used to extract the metal from the ore depends on the position of the metal in the reactivity series.

KEY POINT **If a metal is high in the reactivity series its ores are stable and the metal can be obtained only by electrolysis.**

Metals that are obtained by electrolysis include potassium, sodium, calcium, magnesium and aluminium.

You should be able to predict the method used to extract a metal from its ores, given its position in the reactivity series.

KEY POINT **Metals in the middle of the reactivity series do not form very stable ores and they can be extracted by reduction reactions, often with carbon.**

Examples of metals extracted by reduction are zinc, iron and lead.

KEY POINT **Metals low in the reactivity series, if present in ores, can be extracted simply by heating because the ores are unstable.**

For example, mercury can be extracted by heating cinnabar. A few metals such as gold are found uncombined in the Earth.

Extracting metals by reduction

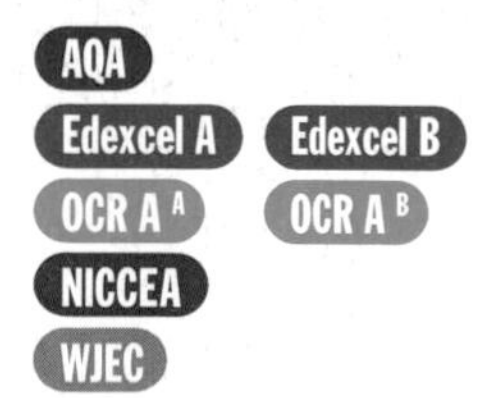

AQA Edexcel A Edexcel B OCR A[A] OCR A[B] NICCEA WJEC

Iron is an example of a metal extracted by **reduction**. The reducing agent is **carbon monoxide**. This removes oxygen from the iron oxide to leave iron.

The extraction of iron is carried out in a **blast furnace** (**Fig. 2.7**).

The furnace is loaded with **iron ore**, **coke** and **limestone** and is heated by blowing hot air into the base from the tuyères. Inside the furnace the following reactions take place:

raises the temperature to about 1500°C:

1. **The burning of the coke in the air:**

$C(s) + O_2(g) \rightarrow CO_2(g)$

carbon + oxygen → carbon dioxide

Fig. 2.7 Blast furnace

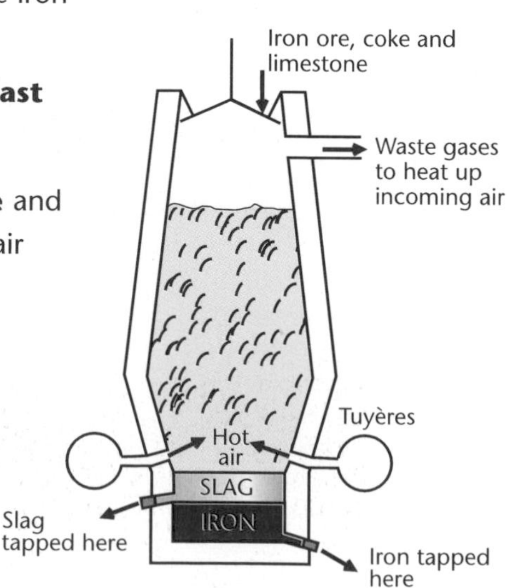

2. **The reduction of the carbon dioxide to carbon monoxide:**

   $CO_2(g) + C(s) \rightarrow 2CO(g)$

   carbon dioxide + carbon → carbon monoxide

This is the important reduction step.

3. **The reduction of the iron ore to iron by carbon monoxide:**

   $Fe_2O_3(s) + 3CO(g) \rightarrow 2Fe(l) + 3CO_2(g)$

   iron(III) oxide + carbon monoxide → iron + carbon dioxide

4. **The decomposition of the limestone produces extra carbon dioxide:**

   $CaCO_3(s) \rightarrow CaO(s) + CO_2(g)$

   calcium carbonate → calcium oxide + carbon dioxide

This step removes impurities from the furnace so it can keep working.

5. **The removal of impurities by the formation of slag:**

   $CaO(s) + SiO_2(s) \rightarrow CaSiO_3(l)$

   calcium oxide + silicon dioxide → calcium silicate (slag)

The molten iron sinks to the bottom of the furnace and the slag floats on the surface of the molten iron. Periodically, the **iron** and **slag** can be tapped off.

The iron produced is called **pig iron** and contains about 4 per cent carbon. Most of this is turned into the alloy called **steel**.

The slag is used as a **phosphorus fertiliser** and for **road building**.

Purifying metals by electrolysis

AQA
Edexcel A Edexcel B
OCR A [A] OCR A [B]
NICCEA
WJEC

KEY POINT

Pure copper is a better electrical conductor than impure copper. It is an economic advantage to purify copper to a high purity.

Copper is purified by **electrolysis** using the cell shown in **Fig. 2.8.**

A pure copper rod (called the **cathode** because it is connected to the negative terminal of the battery) and an impure copper rod (called the **anode**) are used. They are dipped into copper (II) sulphate solution (**electrolyte**).

KEY POINT

During the electrolysis, copper from the anode goes into solution as copper ions and copper ions from the solution are deposited on the cathode.

These changes can be summarised by:

Anode $Cu \rightarrow Cu^{2+} + 2e^-$ (oxidised)

Cathode $Cu^{2+} + 2e^- \rightarrow Cu$ (reduced)

The impurities collect in the anode mud. As copper ores become rare and expensive, much of the new copper needed is obtained by **recycling** old copper wires and pipes.

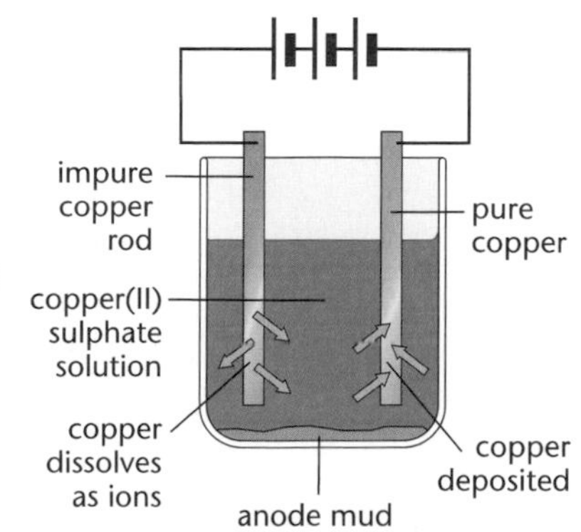

Fig. 2.8 Purification of copper

Extracting metals by electrolysis

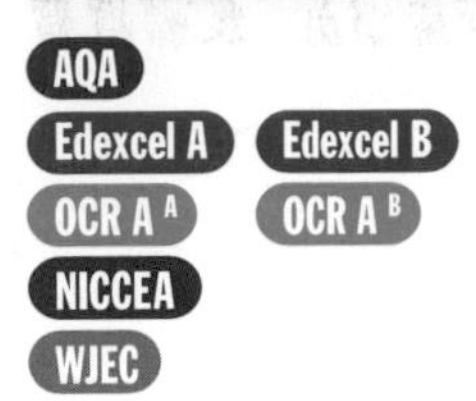

Aluminium is extracted from purified aluminium oxide by electrolysis.

Aluminium has a very high melting point and is not readily soluble in water. It does dissolve in molten cryolite(Na_3AlF_6). A **solution of aluminium oxide in molten cryolite** is a suitable electrolyte. The cell is shown in **Fig. 2.9**.

It takes the same amount of electricity to produce a tonne of aluminium as it does for all of the houses in a small town to use electricity for 1 hour.

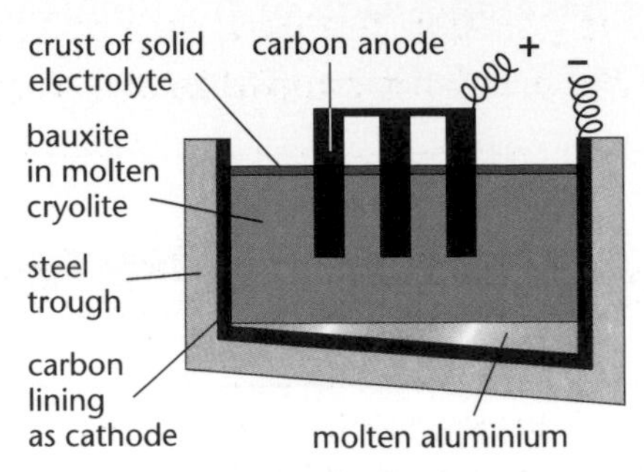

Fig. 2.9 Extraction of aluminium

The electrodes are made of carbon.

In Anglesey in Wales an aluminium smelter uses electricity from the National Grid. A contract is negotiated to ensure electricity at an economic price.

The reactions taking place at the electrodes are:

cathode $Al^{3+} + 3e^- \rightarrow Al$ (reduced)

anode $2O^{2-} \rightarrow O_2 + 4e^-$ (oxidised)

overall reaction:

$4Al^{3+} + 6O^{2-} \rightarrow 4Al + 3O_2$

As this process requires a large amount of electricity, an inexpensive source, e.g. hydroelectric power, is an advantage.

At the working temperature of the cell, the **oxygen** reacts with the **carbon** of the anode to produce **carbon dioxide**. The anode has, therefore, to be replaced frequently.

PROGRESS CHECK

1. What is the name of: a common ore of iron and a common ore of aluminium?
2. What method is used to extract a metal at the top of the reactivity series?
3. Name a metal extracted using carbon monoxide.
4. What is tapped off from a blast furnace used for iron extraction, in addition to molten iron?
5. At which electrode is aluminium produced during its extraction?
6. Write an ionic equation for formation of aluminium from aluminium ions.
7. Why is cryolite used in aluminium extraction?
8. Write down the name of a metal purified by electrolysis.

1. Haematite (or magnetite or iron pyrites), bauxite; 2. Electrolysis; 3. Iron (or zinc);
4. Slag (or calcium silicate); 5. Negative electrode (or cathode); 6. $Al^{3+} + 3e^- \rightarrow Al$;
7. As a solvent for aluminium oxide (or aluminium oxide has a very high melting point);
8. Copper.

2.3 Useful products from air

LEARNING SUMMARY

After studying this section you should be able to:

- ***recall that ammonia is made from nitrogen and hydrogen***
- ***understand the steps in producing ammonia by the Haber process***
- ***recall that a growing plant needs large quantities of nitrogen, phosphorus and potassium and know why these elements are needed***
- ***recall how one nitrogen fertiliser, ammonium nitrate, is made***
- ***understand the problems caused by the over-use of nitrogen fertilisers.***

Ammonia

Ammonia, NH_3, is a compound of nitrogen and hydrogen. It is produced in large quantities by the Haber process using nitrogen from the air as a raw material.

The process is summarised by the diagram below.

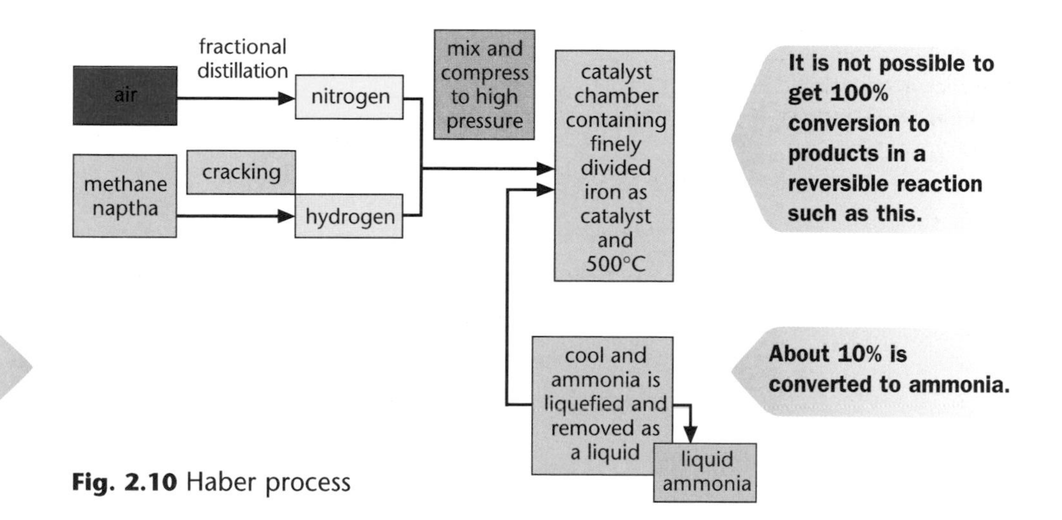

Fig. 2.10 Haber process

It is not possible to get 100% conversion to products in a reversible reaction such as this.

Unreacted nitrogen and hydrogen recycled.

About 10% is converted to ammonia.

The equation for the reaction is:

$$N_2 + 3H_2 \rightleftharpoons 2NH_3$$

The usual arrow between the reactants and the products is replaced a **reversible reaction sign**. This means that the products can decompose, reforming the reactants.

By choosing the best conditions, chemists attempt to produce the highest **yield** of ammonia economically.

The best conditions are:

1. **One part of nitrogen to 3 parts of hydrogen by volume.**
2. **A high pressure.**
3. **A low temperature**. However, using a low temperature reduces the rate of reaction. Using an **iron catalyst** speeds up the reaction.

Catalysts are usually transition metals or transition metal compounds.

In practice, the Haber process operates at a temperature of about 450°C and is used with a catalyst of iron. A high pressure, e.g. 200 atmospheres, is used.

The process is called the **Haber process** after the German scientist, Fritz Haber, who discovered the conditions which would enable ammonia to be made on a large scale.

Nitrogen fertilisers

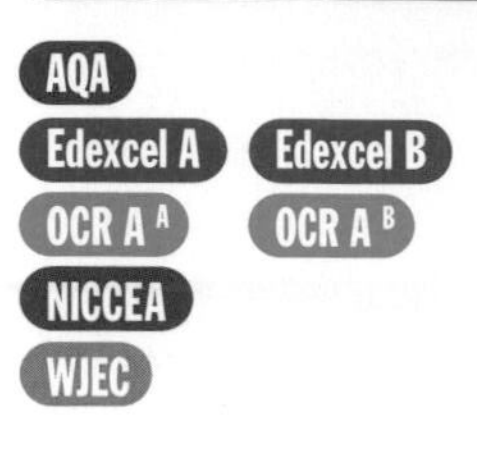

Useful elements in fertilisers

Three elements, **nitrogen**, **phosphorus** and **potassium** are required in large quantities by a healthy plant.

Table 2.5 summarises the importance of these elements to the growing plant and gives some natural and artificial sources.

Table 2.5 Source and use of elements in fertilisers

Element	Importance to growing plant	Natural sources	Artificial sources
Nitrogen	For growth of stems and leaves	Manure, bird droppings, dried blood	Ammonium nitrate, ammonium sulphate, urea
Phosphorus	For root growth	Bone meal	Ammonium phosphate
Potassium	For flowers and fruit	Wood ash	Potassium sulphate

In the past 100 years there has been a **huge growth in population** and therefore in the **quantity of food** required to feed everybody. The development of better and cheaper fertilisers has enabled food production to increase.

Fig. 2.11 summarises how one fertiliser, ammonium nitrate, is made from ammonia produced in the Haber process.

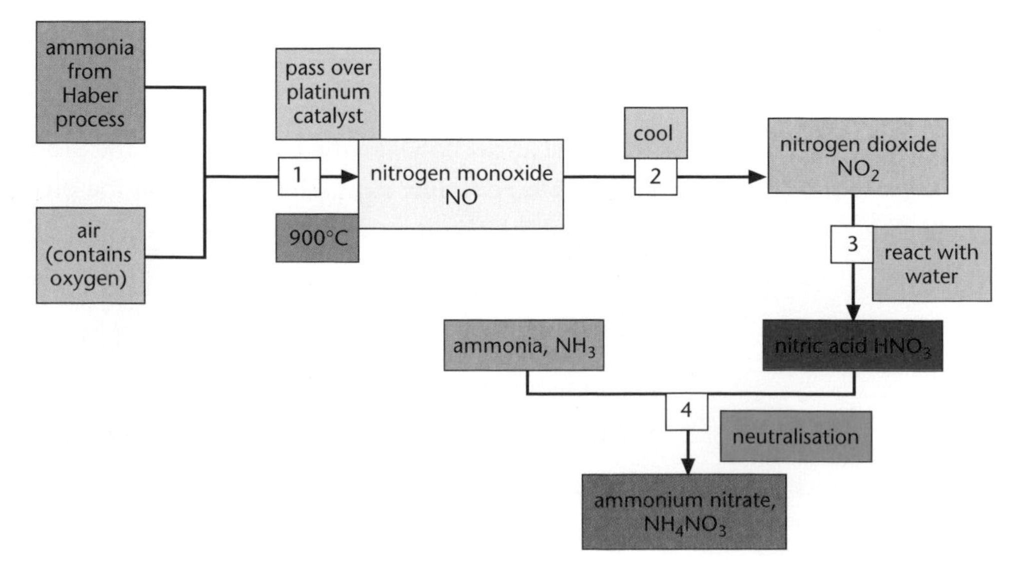

Fig. 2.11 Ammonium nitrate production

The equations for the reactions taking place are:

1. ammonia + oxygen → nitrogen monoxide + water

   $4NH_3(g) + 5O_2(g) \rightarrow 4NO(g) + 6H_2O(g)$

2. nitrogen monoxide + oxygen ⇌ nitrogen dioxide

   $2NO(g) + O_2(g) \rightleftharpoons 2NO_2(g)$

3. nitrogen dioxide + water + oxygen → nitric acid

   $4NO_2(g) + 2H_2O(l) + O_2(g) \rightarrow 4HNO_3(l)$

4. nitric acid + ammonia → ammonium nitrate

   $HNO_3(aq) + NH_3(aq) \rightarrow NH_4NO_3(aq)$

This is one of the hardest equations to balance that you will find.

These equations refer to the steps in the flow diagram on page 36.

Over-use of nitrogen fertilisers

Nitrogen fertilisers in the soil are turned into **nitrates**. These are absorbed into plants in solution through the **roots**.

Nitrates are very soluble and so can be washed out of the soil by rain. If they are washed into rivers a series of changes may take place. This leads to **eutrophication** when there is little life left in the river.

1. Nitrates make water plants grow and these cover the surface of the river.
2. These shade the surface, preventing **light** getting into the water and stopping **photosynthesis**.
3. When these plants die, **bacteria** in the river decompose them.
4. These bacteria use up **oxygen**.
5. There is little oxygen left dissolved in the water, and fish and other life die.

There is a similar result if sewage waste escapes into a river.

Nitrates in water can cause problems to health e.g. blue baby syndrome.

PROGRESS CHECK

1. A fertiliser contains potassium nitrate and ammonium sulphate. Which two elements in the fertiliser are needed in large quantities by growing plants?
2. Which two substances are needed to manufacture ammonium sulphate?
3. What type of reaction is taking place in question 2?
4. Ammonia is a compound of two elements? What are these two elements?
5. What is the meaning of the sign ⇌ in the reaction?
6. What is the catalyst in the Haber process?
7. What is the source of nitrogen and hydrogen in the Haber process?
8. Ammonium nitrate dissolves readily in water. Urea does not dissolve in water but reacts with water slowly to produce ammonia. Suggest when each fertiliser would be used.

1. Nitrogen and potassium; 2. Ammonia and sulphuric acid; 3. Neutralisation; 4. Nitrogen and hydrogen; 5. Reversible reaction; 6. Iron; 7. nitrogen from the air, hydrogen from methane or naphtha; 8. Ammonium nitrate is a quick-acting fertiliser and urea is a slow-acting fertiliser.

2.4 Quantitative chemistry

LEARNING SUMMARY

After studying this section you should be able to:

- ***write chemical formulae and write balanced symbol equations***
- ***calculate quantities of chemicals reacting or produced***
- ***use masses to calculate the simplest formula for a compound***

Equations

AQA
Edexcel A Edexcel B
OCR A [A] OCR A [B]
NICCEA
WJEC

Word equations

A chemical reaction can be summarised by an **equation**.The simplest equation is a **word equation**:

e.g. when **sodium hydroxide** and **hydrochloric acid** are reacted, **sodium chloride** and **water** are formed.

sodium hydroxide + hydrochloric acid → sodium chloride + water

The substances on the left-hand side (sodium hydroxide and hydrochloric acid, in this case) are called **reactants**. The substances produced (sodium chloride and water) are called **products**.

Although word equations may be useful, they do not give a full picture of what is happening.

Symbol equations

Reactions can be summarised using **chemical symbols**. This is a system which is used throughout the world.

The equation for the reaction between sodium hydroxide and hydrochloric acid is written as:

$NaOH + HCl \rightarrow NaCl + H_2O$

This equation is correctly **balanced**, i.e. there are the same number of each type of atom on each side of the equation.

Calcium hydroxide reacts with hydrochloric acid to form calcium chloride and water. The **formula** for calcium hydroxide is not CaOH but $Ca(OH)_2$ (see page 42) and the formula of calcium chloride is not CaCl but $CaCl_2$.

The equation can be written:

$Ca(OH)_2 + HCl \rightarrow CaCl_2 + H_2O$

This equation is **unbalanced** because there are different numbers of atoms on each side:

Notice that the small number after the bracket multiplies everything in the bracket.

Left-hand side		Right-hand side	
1	Ca	1	Ca
2	O	1	O
3	H	2	H
1	Cl	2	Cl

It is an important law, called the **law of conservation of mass**, that atoms cannot be made or destroyed during a chemical reaction, just rearranged.

You cannot write:

$CaOH + HCl \rightarrow CaCl + H_2O$

This would mean altering the formulae and this you cannot do.

Instead, you have to change the proportions of these substances by altering the large numbers at the front:

$Ca(OH)_2 + 2HCl \rightarrow CaCl_2 + 2H_2O$

If you are entering Higher tier, you should be able to write balanced symbol equations.

Left-hand side		Right-hand side	
1	Ca	1	Ca
2	O	2	O
4	H	4	H
2	Cl	2	Cl

Always check an equation is balanced before moving on. Not balancing an equation loses you a mark.

State symbols

Sometimes state symbols are added to symbol equations to show whether the substance is solid, liquid or gas or whether it is in solution.

There is usually no penalty if you miss out state symbols unless the question specifically asks you to add them.

These state symbols are:

(s) solid

(l) liquid

(g) gas

(aq) in aqueous solution, i.e. where the solvent is water.

An example of an equation with state symbols is:

$2Na(s) + 2H_2O(l) \rightarrow 2NaOH(aq) + H_2(g)$

Ionic equations

If we consider the reaction of sodium hydroxide and hydrochloric acid again, sodium hydroxide, hydrochloric acid and sodium chloride are made up of ions.

The equation could be written:

$Na^+ OH^- + H^+ Cl^- \rightarrow Na^+ Cl^- + H_2O$

Ionic equations are usually in questions targeted at A or A*.

Since an equation shows change, anything which appears unchanged on both sides of the equation can be removed.

The simplest equation is

$$OH^- + H^+ \rightarrow H_2O$$

This ionic equation, in addition to having the same number of each type of atom on each side, also has **equal charge on each side**. In this case the sum of the charges on each side is zero.

When you write an ionic equation, check that no ions appear on both sides. These ions called 'spectator ions' can be missed out. An equation shows change.

When chlorine is bubbled through potassium iodide solution, potassium chloride and iodine are produced.

The symbol equation is:

$$2KI + Cl_2 \rightarrow 2KCl + I_2$$

Potassium iodide and potassium chloride are made up of ions.

$$2K^+ I^- + Cl_2 \rightarrow 2K^+ Cl^- + I_2$$

The ionic equation is

$$2\, I^- + Cl_2 \rightarrow 2\, Cl^- + I_2$$

Now there are two iodines and two chlorines on each side and the charge on each side is –2.

PROGRESS CHECK

Balance each of the following equations.

1. $Mg + HCl \rightarrow MgCl_2 + H_2$
2. $Na + Cl_2 \rightarrow NaCl$
3. $H_2 + O_2 \rightarrow H_2O$
4. $H_2O_2 \rightarrow H_2O + O_2$
5. $H_2 + Cl_2 \rightarrow HCl$
6. $NO + O_2 \rightarrow NO_2$
7. $O^{2-} + H^+ \rightarrow H_2O$
8. $Na + H^+ \rightarrow Na^+ + H_2$

1. $Mg + \mathbf{2}HCl \rightarrow MgCl_2 + H_2$; 2. $\mathbf{2}Na + Cl_2 \rightarrow \mathbf{2}NaCl$;
3. $\mathbf{2}H_2 + O_2 \rightarrow \mathbf{2}H_2O$; 4. $\mathbf{2}H_2O_2 \rightarrow \mathbf{2}H_2O + O_2$
5. $H_2 + Cl_2 \rightarrow \mathbf{2}HCl$; 6. $\mathbf{2}NO + O_2 \rightarrow \mathbf{2}NO_2$
7. $O^{2-} + \mathbf{2}H^+ \rightarrow H_2O$; 8. $\mathbf{2}Na + \mathbf{2}H^+ \rightarrow \mathbf{2}Na^+ + H_2$

Relative atomic mass and relative formula mass

AQA Edexcel A Edexcel B OCR A[A] OCR A[B] NICCEA WJEC

Atoms are too small to be weighed individually. It is possible, however, to compare the mass of one atom with the mass of another.

This is done using a **mass spectrometer**. For example, a magnesium atom has twice the mass of a carbon-12 atom and six times the mass of a helium atom.

The relative atomic mass of an atom is the number of times an atom is heavier than one-twelfth of a carbon-12 atom.

Relative atomic masses are not all whole numbers because of the existence of isotopes (see 1.1).

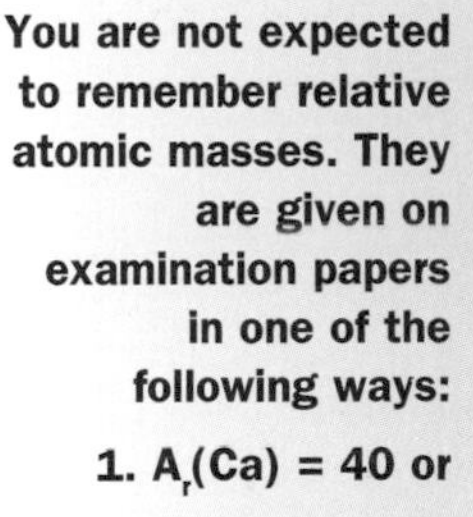

You are not expected to remember relative atomic masses. They are given on examination papers in one of the following ways:

1. $A_r(Ca) = 40$ or
2. Ca = 40

The **relative atomic mass (Ar)** is simply a **number** and has **no units.**

For compounds, if you know the formula, you can use relative atomic masses to work out **relative formula masses (Mr)** , e.g. Work out the relative formula mass of water, H_2O, given $A_r(H) = 1$ and $A_r(O) = 16$

The relative formula mass of water
$M_r = (2 \times 1) + 16 = 18$

The relative formula mass can be worked out by adding relative atomic masses.

Using equations to calculate masses

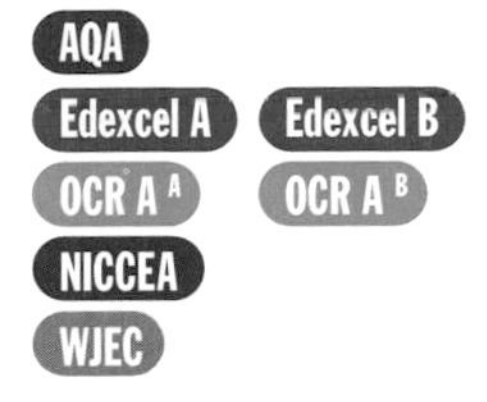

A balanced symbol equation can tell you about the chemicals involved in a reaction as reactants or products. It can also tell you about the **masses** of chemicals which react or are formed.

In order to do this, you need **relative atomic masses** of the elements.

The relative atomic masses can be looked up in a data book or obtained from the periodic table.

Iron and sulphur react together to form iron(II) sulphide

The symbol equation is:

$Fe + S \rightarrow FeS$

Notice the sum of the masses of the reactants equals the mass of the product.

The relative atomic mass of iron is 56 and sulphur, 32

From the equation, using relative atomic masses

56 g of iron combine with 32 g of sulphur to form 88 g of iron(II) sulphide.

Another example:

Carbon burns in excess oxygen to form carbon dioxide.

Calculate the mass of carbon dioxide produced when 1 g of carbon is burned. ($A_r(C) = 12$, $A_r(O) = 16$)

First write the symbol equation:

$C + O_2 \rightarrow CO_2$

Always check that the mass of the reactants is the same as the mass of the products.

Now use relative atomic masses to work out masses of reactants and products.

12 g of carbon react with (2×16)g of oxygen to form $(12 + (2 \times 16\ g)$ of carbon dioxide

12 g of carbon react with 32 g of oxygen to form 44 g of carbon dioxide

If 1 g of carbon is used (one-twelfth of the quantity), the mass of carbon dioxide formed would be one-twelfth, i.e. $44/12 = 3.7$ g

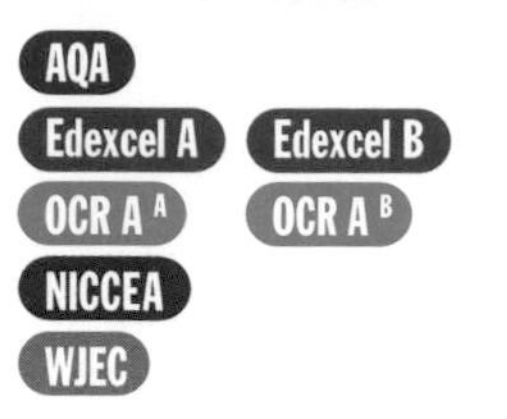

($A_r(H) = 1$, $A_r(C) = 12$, $A_r(O) = 16$, $A_r(Mg) = 24$, $A_r(S) = 32$, $A_r(K) = 39$)

1. How many times heavier is a magnesium atom than a carbon atom?
2. What is the relative formula mass of methane, CH_4?
3. What is relative formula mass of sulphuric acid, H_2SO_4?

The equation for the action of heat on potassium hydrogencarbonate

$2KHCO_3 \rightarrow K_2CO_3 + H_2O + CO_2$

4. What is the relative formula mass of potassium hydrogencarbonate?
5. What is the relative formula mass of potassium carbonate?
6. What mass of potassium carbonate would be formed when 10 g of potassium hydrogencarbonate are completely decomposed?

1. Twice; 2. 16; 3. 98; 4. 100; 5. 138; 6. 13.8 g

Working out chemical formulae

AQA · Edexcel A · Edexcel B · OCR A [A] · OCR A [B] · NICCEA · WJEC

Chemical formulae can be worked out using the formulae of common ions.

Table 2.6 contains some of the common positive and negative ions.

Positive ions			Negative ions		
+1	+2	+3	−1	−2	−3
sodium Na^+	magnesium Mg^{2+}	aluminium Al^{3+}	chloride Cl^-	sulphate SO_4^{2-}	phosphate PO_4^{3-}
potassium K^+	calcium Ca^{2+}		nitrate NO_3^-	carbonate CO_3^{2-}	
hydrogen H^+	lead Pb^{2+}		hydroxide OH^-	oxide O^{2-}	
ammonium NH_4^+	zinc Zn^{2+}				
silver Ag^+	copper Cu^{2+}				

If you want to write a chemical formula, you will need to use the correct ions:

Remember metals form positive ions.

e.g. **Sodium chloride** Na^+ and Cl^-

As there are **equal numbers of positive and negative charges**, you can write the formula as NaCl.

Sodium sulphate Na^+ and SO_4^{2-}

There are **twice as many negative charges as positive charges**.

In the formula there need to be twice as many sodium ions.

The formula is therefore written as Na_2SO_4.

Aluminium oxide Al^{3+} and O^{2-}

If you are taking Higher tier, you should be able to recognise incorrect formulae and write correct ones.

In order to get equal numbers of positive and negative charges, you have to take two aluminium ions for every three oxide ions. The formula is Al_2O_3.

It is possible to work out the formula of a compound using results from an experiment.

Magnesium oxide

If magnesium burns in oxygen, magnesium oxide is formed.

Here are the results of an experiment.

Mass of crucible + lid = 25.15 g

Mass of crucible + lid + magnesium = 25.27 g

∴ Mass of magnesium = 25.27 – 25.15 = 0.12 g

Mass of crucible + lid + magnesium oxide = 25.35 g

∴ Mass of magnesium oxide = 25.35 – 25.15 g = 0.20 g

From these results

0.12 g of magnesium combines with (0.20 – 0.12 g) of oxygen to form 0.20 g of magnesium oxide.

0.12 g of magnesium combines with 0.08 g of oxygen.

Divide each mass by the appropriate relative atomic mass:
$A_r(Mg) = 24$, $A_r(O) = 16$.

	Magnesium	Oxygen
	$\frac{0.12}{24}$	$\frac{0.08}{16}$
	= 0.05	= 0.05

Divide by the smallest number (here they are both the same)

	Magnesium	Oxygen
	1	1

The simplest formula is MgO.

Lead oxide

4.14 g of lead combines with 0.64 g of oxygen

$A_r(Pb) = 207$, $A_r(O) = 16$

Divide by the appropriate relative atomic masses

	Lead	Oxygen
	$\frac{4.14}{207}$	$\frac{0.64}{16}$
	0.02	0.04

Divide by the smallest, i.e. 0.02

	Lead	Oxygen
	1	2

A common mistake here is to write the formula as Pb_2O.

The simplest formula for this lead oxide is PbO_2.

PROGRESS CHECK

$A_r(H) = 1$, $A_r(C) = 12$, $A_r(N) = 14$, $A_r(O) = 16$, $A_r(Cu) = 64$
60 g of copper oxide produces 1.28 g of copper.

1. What mass of oxygen combines with 1.28 g of copper?
2. Choose the formula of this copper oxide from the list:
   Cu_2O CuO CuO_2
3. 6 g of carbon combines with 1 g of hydrogen.
   Choose the **simplest** formula of this compound.
   CH_2 C_2H_4 CH_4
   0.7 g of nitrogen combines to form 1.5 g of nitrogen oxide.
4. What mass of oxygen combines with 0.7 g of nitrogen?
5. Choose the simplest formula of this compound.
   N_2O NO_2 NO

1. 0.32 g; 2. CuO; 3. CH_2; 4. 0.8g; 5. NO

2.5 Earth cycles

LEARNING SUMMARY

After studying this section you should be able to:

- *recall the approximate percentages of the gases in a typical sample of dry air*
- *understand how the composition of the atmosphere has changed over history*
- *understand the role of the oceans in maintaining the composition of the atmosphere*
- *understand the rock record.*

Changes in composition of atmosphere and oceans

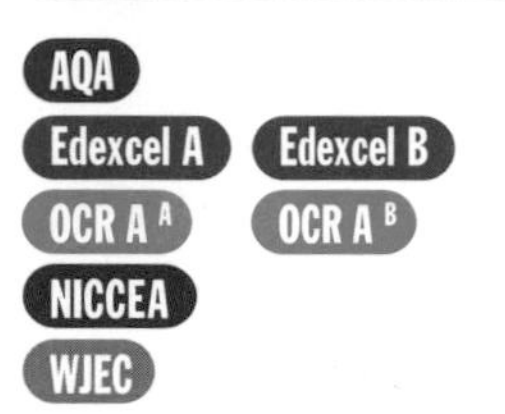

Composition of the present atmosphere

Air is a **mixture** of gases. Its composition can vary from place to place.

The typical composition of a sample of dry air is

Nitrogen	78%
Oxygen	21%
Argon (and other noble gases)	1%
Carbon dioxide	0.04%

Students frequently write that air contains hydrogen. Normal air does not.

The apparatus in **Fig. 2.12** can be used to find the percentage of oxygen (the active gas) in air.

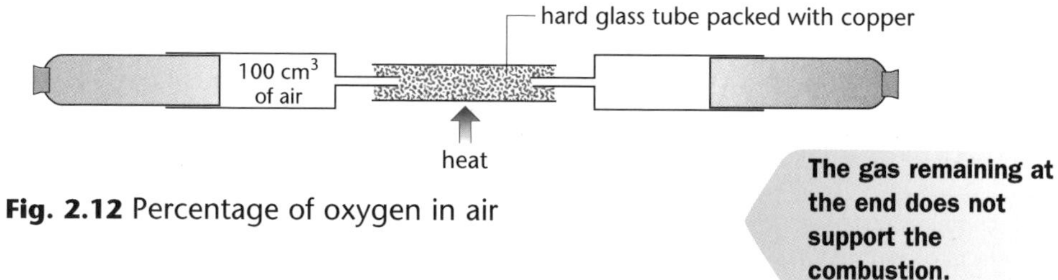

Fig. 2.12 Percentage of oxygen in air

The gas remaining at the end does not support the combustion.

A sample of air is passed backwards and forwards over heated **copper**. The **oxygen** in the air is removed. Black **copper(II) oxide** is formed.

$2Cu + O_2 \rightarrow 2\ CuO$

It is important all volumes are measured at room temperature. At higher temperature gases would be expanded.

The percentage of oxygen in the air can be calculated by measuring the volume remaining when the apparatus has cooled to room temperature.

Page 45 is not required for NICCEA Chemistry.

How the atmosphere has changed

Table 2.7 Effects of changes on the atmosphere

Change	Effect on the atmosphere
The first atmosphere	Consisted mainly of **hydrogen** and **helium**
Volcanoes started to erupt	Mostly **carbon dioxide** and **water vapour** entering the atmosphere. Smaller quantities of **methane** and **ammonia**
Earth cools	Water vapour condenses to **liquid water.** Oceans started to form
Nitrifying and denitrifying bacteria start to work	Ammonia is converted into **nitrates,** and **nitrates** are converted into gaseous **nitrogen**
Methane in the atmosphere burns	**Carbon dioxide** is formed
Photosynthesis occurs	Plants convert carbon dioxide into **oxygen**
Increasing levels of carbon dioxide	Due to burning of fossil fuels and destruction of environment reducing photosynthesis

Oceans

Oceans were formed when the Earth cooled down and the condensed water settled in the lowest points on the Earth's surface. Water is a **solvent** and started to **dissolve substances** from the **rocks of the Earth**. Over hundreds of millions of years the composition of the sea has become fairly constant over the whole globe. Although the actual concentration of ions varies from place to place, the **ratio of each ion** with respect to another is **fairly constant**.

Some rocks contain minerals the do not dissolve but are broken down by chemical weathering. The products dissolve – e.g. calcium carbonate is turned into calcium hydrogencarbonate.

There is a balance in the oceans, keeping the concentrations of dissolved ions approximately constant.

Rainwater dissolves minerals as it filters through rocks and into rivers. These ions are washed into the seas.

There are **three ways in which these ions are removed.**

1. If the concentrations of the ions becomes too great, ions will **precipitate**.

   e.g. if the concentrations of calcium and sulphate ions become too great, calcium sulphate is precipitated.

   $Ca^{2+} (aq) + SO_4^{2-} (aq) \rightarrow CaSO_4(s)$

2. Marine animals called molluscs have **shells made of calcium carbonate**.

   These shells are formed by taking calcium and carbonate ions from the water.

   $Ca^{2+} (aq) + CO_3^{2-}(aq) \rightarrow CaCO_3$

3. In some countries, **sea water is evaporated** to produce sea salt which is then sold.

The Dead Sea has higher levels of dissolved ions. Rivers running into the Dead Sea contain dissolved minerals that are concentrated when water evaporates.

The most common ions in sea water are **sodium** and **chloride**, but there are small concentrations of many other ions. Magnesium and bromine are extracted from sea water.

The oceans also contain **phytoplankton**. These are a large form of plankton, which undergo **photosynthesis**, removing carbon dioxide and replacing it with oxygen. It has been estimated that phytoplankton are more important than rain forests for removing carbon dioxide and replacing it with oxygen.

It is suggested that rising ocean temperatures will cause more phytoplankton to be produced. This could remove more carbon dioxide. Pollution, however, could reduce levels of phytoplankton.

Carbon cycle

AQA
Edexcel A Edexcel B
OCR A[A] OCR A[B]

NICCEA
WJEC

KEY POINT

The concentration of carbon dioxide has been kept constant by a delicate balance between respiration and combustion on one side and photosynthesis on the other.

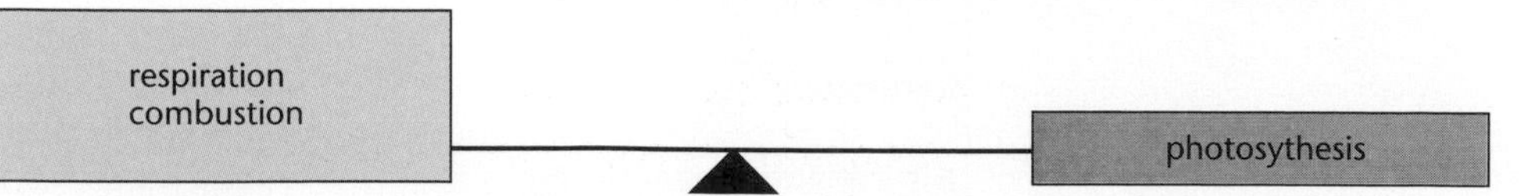

Fig. 2.13 Balance between respiration, combustion and photosynthesis

Over the last 50 years, this balance has been disturbed. The main reason for this is the **increased burning of fossil fuels** producing carbon dioxide and the destruction of forests and green plants which would remove carbon dioxide.

The increasing **concentration of carbon dioxide** (and other greenhouse gases) has caused a rise in the average temperature of the Earth's atmosphere called **global warming**.

Fig. 2.14 shows how the **greenhouse effect** brings about global warming.

One hectare (2.47 acres) of Sitka Spruce trees in a year uses up 6 tonnes of carbon dioxide, producing four tonnes of oxygen.

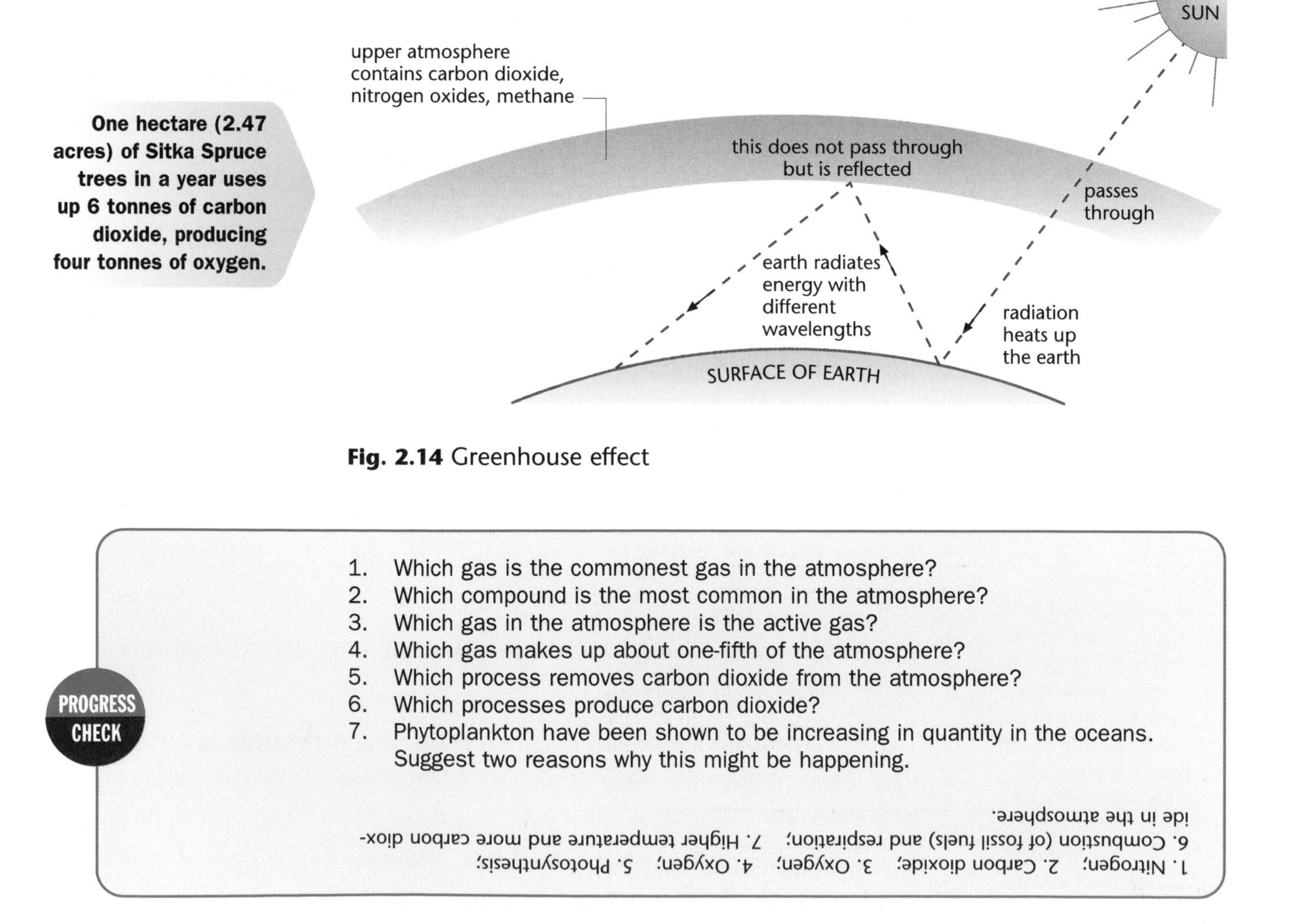

Fig. 2.14 Greenhouse effect

PROGRESS CHECK

1. Which gas is the commonest gas in the atmosphere?
2. Which compound is the most common in the atmosphere?
3. Which gas in the atmosphere is the active gas?
4. Which gas makes up about one-fifth of the atmosphere?
5. Which process removes carbon dioxide from the atmosphere?
6. Which processes produce carbon dioxide?
7. Phytoplankton have been shown to be increasing in quantity in the oceans. Suggest two reasons why this might be happening.

1. Nitrogen; 2. Carbon dioxide; 3. Oxygen; 4. Oxygen; 5. Photosynthesis;
6. Combustion (of fossil fuels) and respiration; 7. Higher temperature and more carbon dioxide in the atmosphere.

The rock record

AQA
Edexcel A Edexcel B
OCR A A OCR A B
WJEC

The rock record represents the evidence that is left behind from events in the past that have led to the formation and distortion of layers of rocks. These layers form a **succession**. Rocks in succession usually become **progressively older, moving down**.

Rock successions may be **faulted**, **folded** or may contain **fossils**. All of these may provide evidence to help date rocks and suggest what may have happened to them in the past.

Faults and folds

When large plates of the Earth move closer together, the layers of rock may **fold** or may be broken (**fault**). These are shown in **Fig. 2.15**.

Geologists can use information about faults and folds to work out the order in which events have happened.

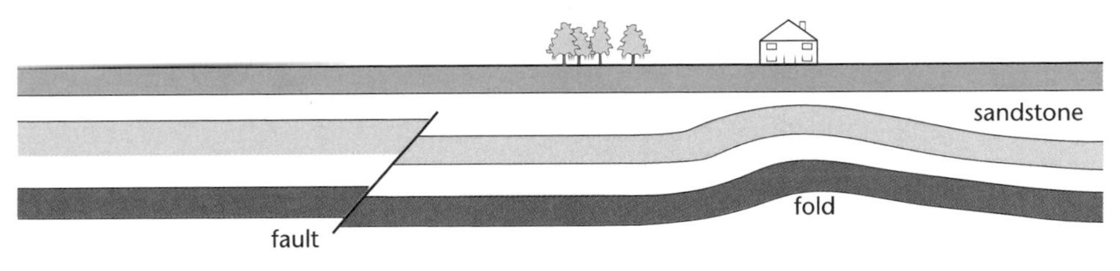

Fig. 2.15 Faults and folds

In **Fig. 2.15** the fault must have occurred after the fold, as the sandstone layer has been affected by the fault but not by the fold.

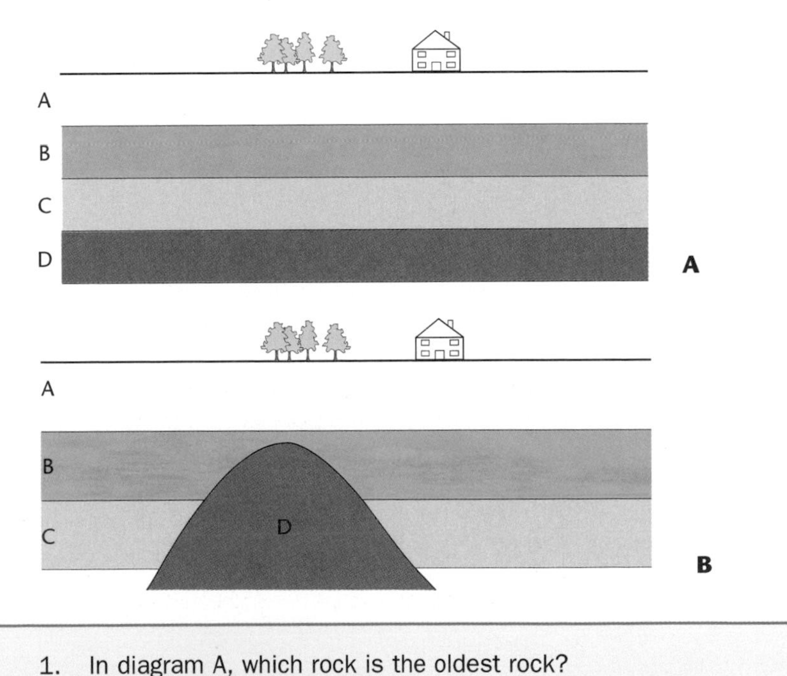

PROGRESS CHECK

1. In diagram A, which rock is the oldest rock?
   Look at diagram B
2. Which rock is an igneous rock?
3. How do you know that rock D is older than rock A?

1. D; 2. D; 3. The rock D does not cut into layer A.

Sample GCSE questions

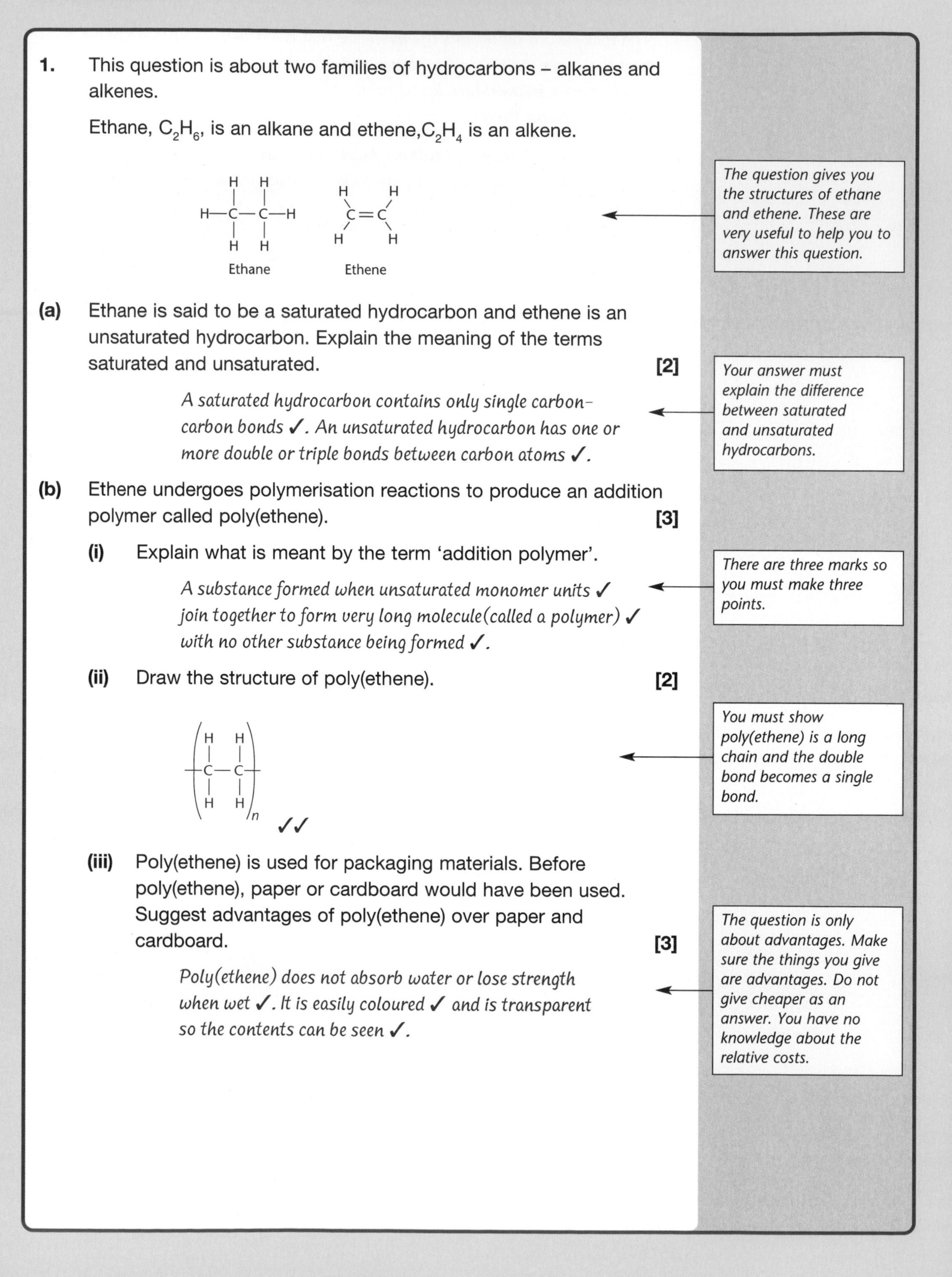

1. This question is about two families of hydrocarbons – alkanes and alkenes.

Ethane, C_2H_6, is an alkane and ethene, C_2H_4 is an alkene.

Ethane   Ethene

The question gives you the structures of ethane and ethene. These are very useful to help you to answer this question.

(a) Ethane is said to be a saturated hydrocarbon and ethene is an unsaturated hydrocarbon. Explain the meaning of the terms saturated and unsaturated. **[2]**

A saturated hydrocarbon contains only single carbon-carbon bonds ✓. An unsaturated hydrocarbon has one or more double or triple bonds between carbon atoms ✓.

Your answer must explain the difference between saturated and unsaturated hydrocarbons.

(b) Ethene undergoes polymerisation reactions to produce an addition polymer called poly(ethene). **[3]**

(i) Explain what is meant by the term 'addition polymer'.

A substance formed when unsaturated monomer units ✓ join together to form very long molecule (called a polymer) ✓ with no other substance being formed ✓.

There are three marks so you must make three points.

(ii) Draw the structure of poly(ethene). **[2]**

✓✓

You must show poly(ethene) is a long chain and the double bond becomes a single bond.

(iii) Poly(ethene) is used for packaging materials. Before poly(ethene), paper or cardboard would have been used. Suggest advantages of poly(ethene) over paper and cardboard. **[3]**

Poly(ethene) does not absorb water or lose strength when wet ✓. It is easily coloured ✓ and is transparent so the contents can be seen ✓.

The question is only about advantages. Make sure the things you give are advantages. Do not give cheaper as an answer. You have no knowledge about the relative costs.

Sample GCSE questions

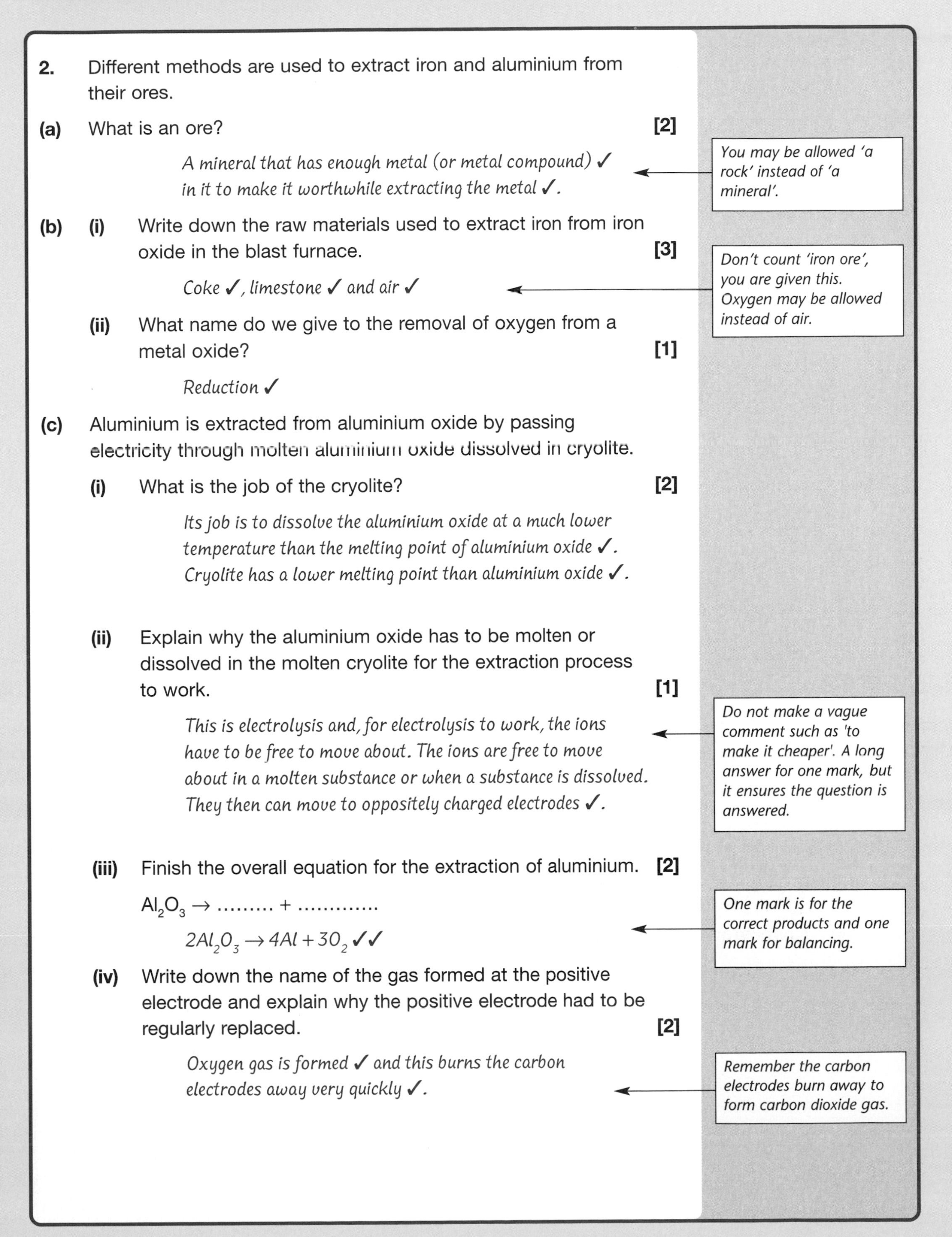

2. Different methods are used to extract iron and aluminium from their ores.

(a) What is an ore? **[2]**

A mineral that has enough metal (or metal compound) ✓ in it to make it worthwhile extracting the metal ✓.

> *You may be allowed 'a rock' instead of 'a mineral'.*

(b) (i) Write down the raw materials used to extract iron from iron oxide in the blast furnace. **[3]**

Coke ✓, limestone ✓ and air ✓

> *Don't count 'iron ore', you are given this. Oxygen may be allowed instead of air.*

(ii) What name do we give to the removal of oxygen from a metal oxide? **[1]**

Reduction ✓

(c) Aluminium is extracted from aluminium oxide by passing electricity through molten aluminium oxide dissolved in cryolite.

(i) What is the job of the cryolite? **[2]**

Its job is to dissolve the aluminium oxide at a much lower temperature than the melting point of aluminium oxide ✓. Cryolite has a lower melting point than aluminium oxide ✓.

(ii) Explain why the aluminium oxide has to be molten or dissolved in the molten cryolite for the extraction process to work. **[1]**

This is electrolysis and, for electrolysis to work, the ions have to be free to move about. The ions are free to move about in a molten substance or when a substance is dissolved. They then can move to oppositely charged electrodes ✓.

> *Do not make a vague comment such as 'to make it cheaper'. A long answer for one mark, but it ensures the question is answered.*

(iii) Finish the overall equation for the extraction of aluminium. **[2]**

$Al_2O_3 \rightarrow$ ……… + ………….

$2Al_2O_3 \rightarrow 4Al + 3O_2$ ✓✓

> *One mark is for the correct products and one mark for balancing.*

(iv) Write down the name of the gas formed at the positive electrode and explain why the positive electrode had to be regularly replaced. **[2]**

Oxygen gas is formed ✓ and this burns the carbon electrodes away very quickly ✓.

> *Remember the carbon electrodes burn away to form carbon dioxide gas.*

Sample GCSE questions

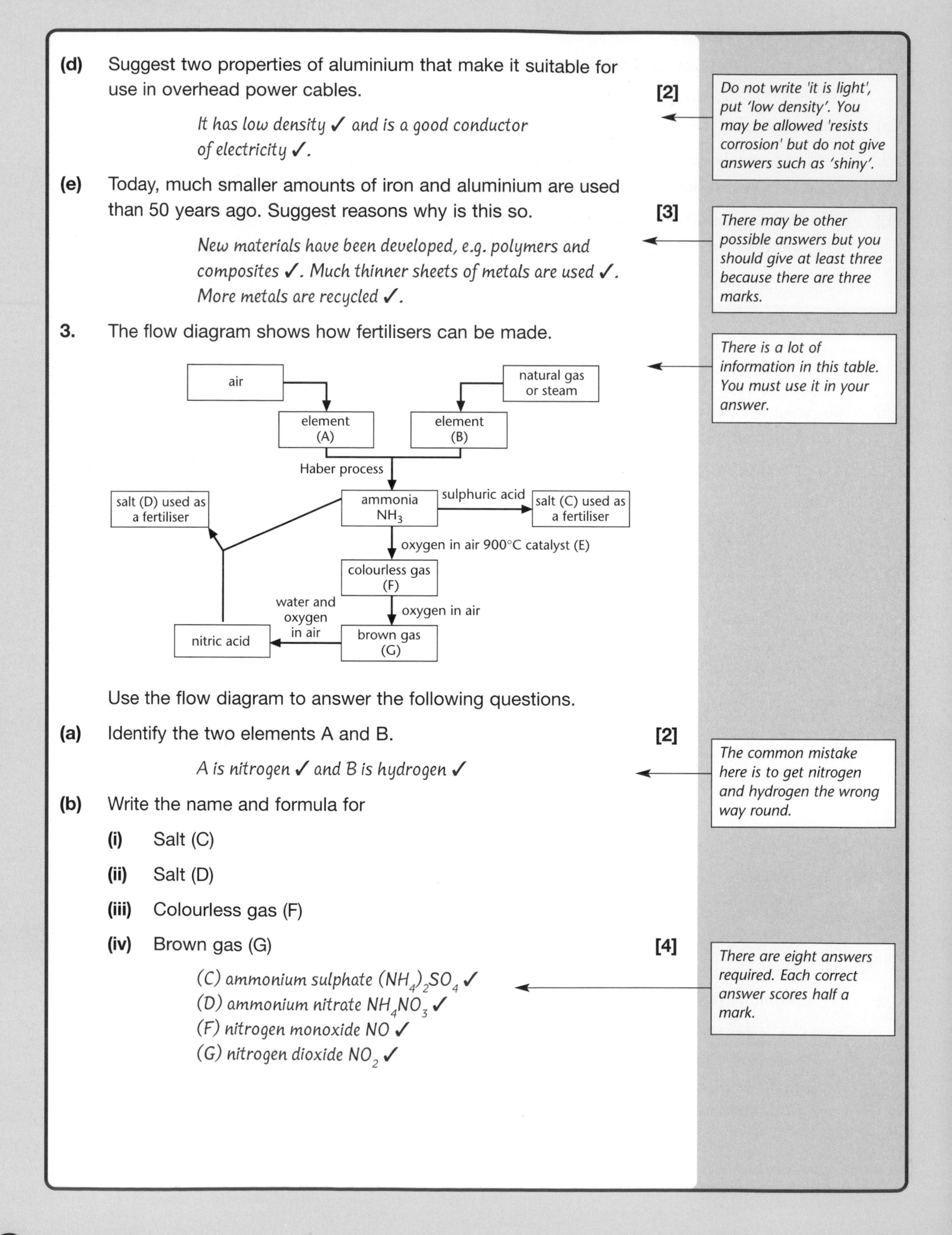

(d) Suggest two properties of aluminium that make it suitable for use in overhead power cables. **[2]**

It has low density ✓ and is a good conductor of electricity ✓.

> *Do not write 'it is light', put 'low density'. You may be allowed 'resists corrosion' but do not give answers such as 'shiny'.*

(e) Today, much smaller amounts of iron and aluminium are used than 50 years ago. Suggest reasons why is this so. **[3]**

New materials have been developed, e.g. polymers and composites ✓. Much thinner sheets of metals are used ✓. More metals are recycled ✓.

> *There may be other possible answers but you should give at least three because there are three marks.*

3. The flow diagram shows how fertilisers can be made.

> *There is a lot of information in this table. You must use it in your answer.*

Use the flow diagram to answer the following questions.

(a) Identify the two elements A and B. **[2]**

A is nitrogen ✓ and B is hydrogen ✓

> *The common mistake here is to get nitrogen and hydrogen the wrong way round.*

(b) Write the name and formula for

(i) Salt (C)

(ii) Salt (D)

(iii) Colourless gas (F)

(iv) Brown gas (G) **[4]**

(C) ammonium sulphate $(NH_4)_2SO_4$ ✓
(D) ammonium nitrate NH_4NO_3 ✓
(F) nitrogen monoxide NO ✓
(G) nitrogen dioxide NO_2 ✓

> *There are eight answers required. Each correct answer scores half a mark.*

Sample GCSE questions

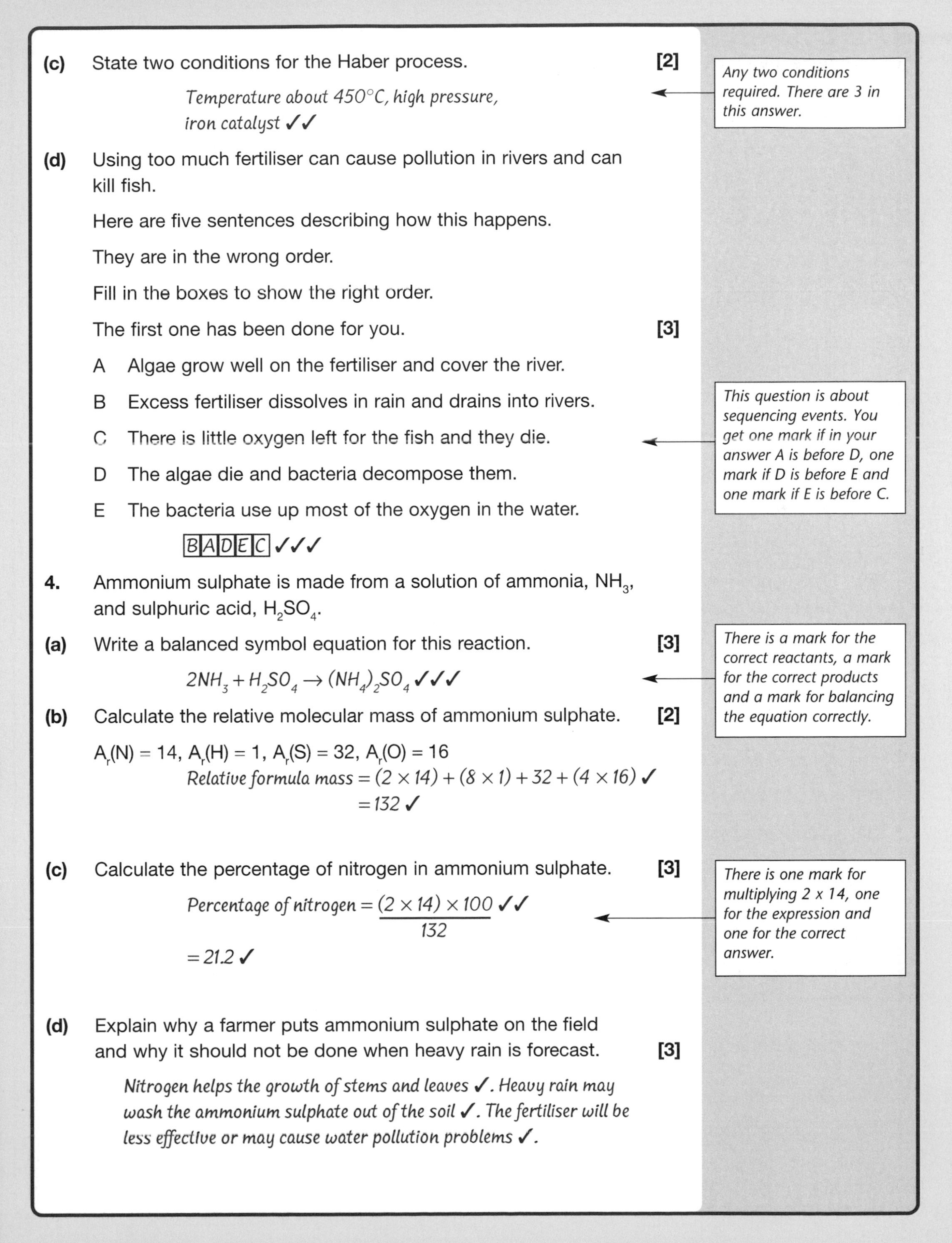

(c) State two conditions for the Haber process. **[2]**

Temperature about 450°C, high pressure, iron catalyst ✓✓

Any two conditions required. There are 3 in this answer.

(d) Using too much fertiliser can cause pollution in rivers and can kill fish.

Here are five sentences describing how this happens.

They are in the wrong order.

Fill in the boxes to show the right order.

The first one has been done for you. **[3]**

A Algae grow well on the fertiliser and cover the river.

B Excess fertiliser dissolves in rain and drains into rivers.

C There is little oxygen left for the fish and they die.

D The algae die and bacteria decompose them.

E The bacteria use up most of the oxygen in the water.

B	A	D	E	C

✓✓✓

This question is about sequencing events. You get one mark if in your answer A is before D, one mark if D is before E and one mark if E is before C.

4. Ammonium sulphate is made from a solution of ammonia, NH_3, and sulphuric acid, H_2SO_4.

(a) Write a balanced symbol equation for this reaction. **[3]**

$2NH_3 + H_2SO_4 \rightarrow (NH_4)_2SO_4$ ✓✓✓

There is a mark for the correct reactants, a mark for the correct products and a mark for balancing the equation correctly.

(b) Calculate the relative molecular mass of ammonium sulphate. **[2]**

$A_r(N) = 14$, $A_r(H) = 1$, $A_r(S) = 32$, $A_r(O) = 16$

Relative formula mass $= (2 \times 14) + (8 \times 1) + 32 + (4 \times 16)$ ✓

$= 132$ ✓

(c) Calculate the percentage of nitrogen in ammonium sulphate. **[3]**

Percentage of nitrogen $= \dfrac{(2 \times 14) \times 100}{132}$ ✓✓

$= 21.2$ ✓

There is one mark for multiplying 2 x 14, one for the expression and one for the correct answer.

(d) Explain why a farmer puts ammonium sulphate on the field and why it should not be done when heavy rain is forecast. **[3]**

Nitrogen helps the growth of stems and leaves ✓. Heavy rain may wash the ammonium sulphate out of the soil ✓. The fertiliser will be less effective or may cause water pollution problems ✓.

Exam practice questions

1. An experiment was carried out to investigate the rate of reaction between magnesium and sulphuric acid.

0.07 g of magnesium ribbon were reacted with excess dilute sulphuric acid.
The volume of gas produced was recorded every 5 seconds.
The results are shown in the table below.

Time in s	Volume in cm^3	Time in s	Volume in cm^3
0	9	25	63
5	18	30	67
10	34	35	69
15	47	40	70
20	57	45	70

(a) On a piece of graph paper, plot these results with the volume of gas on the y-axis. Draw a smooth curve through the points. **[3]**

(b) When is the reaction fastest? **[1]**

(c) How long does it take for 0.07 g of magnesium to react completely? **[1]**

(d) At what time was 0.02 g of magnesium left unreacted? **[1]**

(e) The experiment was repeated using 0.07 g of magnesium powder instead of magnesium ribbon. How would the graph for this reaction compare with the graph you drew? Explain your answers. **[4]**

2. A sample of copper bromide, CuBr, weighing 21.6 g was heated with excess iron powder. A reaction took place producing copper and iron(III) bromide.

(a) Write a balanced symbol equation for the reaction. **[3]**

(b) What type of reaction took place? **[1]**

(c) Suggest a method of extracting copper from the mixture remaining. **[3]**

(d) Calculate the mass of copper produced – Ar (Cu) = 64, Ar (Br) = 80 **[3]**

Chapter 3

Patterns of behaviour

The following topics are covered in this section:

- ***The periodic table***
- ***Chemical reactions***
- ***Rates of reaction***
- ***Reversible reactions***
- ***Energy transfer in reactions***

What you should know already

Complete the passage using words from the list. You can use words more than once.

acidic	**alkaline**	**calcium**	**carbon dioxide**
chlorine	**copper**	**hydrogen**	**lead**
left	**magnesium**	**magnesium oxide**	**metals**
neutral	**neutralisation**	**nitrogen**	**oxygen**
periodic table	**reactivity series**	**right**	**sodium hydroxide**
symbol	**zinc sulphate**	**universal indicator**	

1. Each element can be represented by one or two letters called a 1.__________.

2. Which element is represented by each of the following:

Ca 2.__________. O 3.__________. Mg 4.__________. Cl 5.__________. Pb 6.__________. N 7.__________.

3. Elements are shown in the 8.__________.

4. Most of the known elements are 9.__________.

5. Metals are on the 10.__________. hand side and non metals on the 11.__________ hand side.

Metals react with oxygen, water, acids and oxides of other metals.

Complete the following word equations:

6. Magnesium + oxygen → 12.__________.

7. Sodium + water → 13.__________ + 14.__________.

8. Zinc + sulphuric acid → 15.__________ + 16.__________.

9. Magnesium + copper(II) oxide → 17.__________ + 18.__________.

10. Metals are arranged in order of reactivity in the 19.__________.

This can be used to predict reactions.

11. The pH of a solution can be found using a pH meter or 20.__________.

12. A solution with a pH of 7 is 21.__________. Solutions with a pH less than 7 is 22.__________ and greater than 7 is 23.__________.

13. Reactions between acids and alkalis are 24.__________ reactions.

ANSWERS

1. symbol; 2.calcium; 3. oxygen; 4.magnesium; 5. chlorine; 6. lead; 7. nitrogen; 8. periodic table; 9. metals; 10. left; 11. right; 12. magnesium oxide; 13 and 14. sodium hydroxide and hydrogen; 15 and 16. zinc sulphate and hydrogen; 17. and 18. magnesium oxide and copper; 19. reactivity series; 20. universal indicator; 21. neutral; 22. acidic; 23. alkaline; 24. neutralisation.

3.1 The periodic table

LEARNING SUMMARY

After studying this section you should be able to:

- ***understand the relationship between the position of an element in the periodic table and the properties of the element***
- ***explain the relationship between the position of an element in the periodic table and the arrangement of electrons in the atoms***
- ***understand the patterns within families of elements: alkali metals, halogens and noble gases***
- ***recall some of the typical properties of transition metals.***

Structure of the periodic table

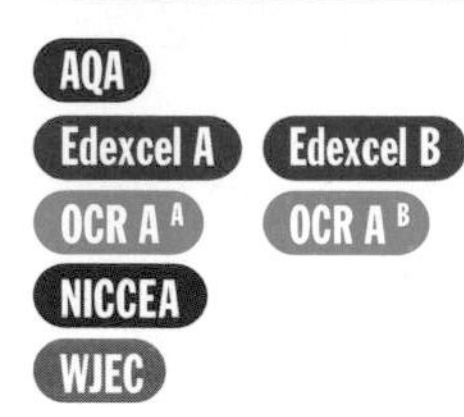

The periodic table is an arrangement of all of the hundred plus elements in order of increasing atomic number, with elements with similar properties in the same vertical column.

The periodic table is shown in its modern form in **Fig. 3.1**.

	1	2											3	4	5	6	7	0
1	1 **H** Hydrogen 1																	4 **He** Helium 2
2	7 **Li** Lithium 3	9 **Be** Berylium 4											11 **B** Boron 5	12 **C** Carbon 6	14 **N** Nitrogen 7	16 **O** Oxygen 8	19 **F** Fluorine 9	20 **Ne** Neon 10
3	23 **Na** Sodium 11	24 **Mg** Magnesium 12	Transition elements										27 **Al** Aluminium 13	28 **Si** Silicon 14	31 **P** Phosphorus 15	32 **S** Sulphur 16	35.5 **Cl** Chlorine 17	40 **Ar** Argon 18
4	39 **K** Potassium 19	40 **Ca** Calcium 20	45 **Sc** Scandium 21	48 **Ti** Titanium 22	51 **V** Vanadium 23	52 **Cr** Chromium 24	55 **Mn** Manganese 25	56 **Fe** Iron 26	59 **Co** Cobalt 27	59 **Ni** Nickel 28	64 **Cu** Copper 29	65 **Zn** Zinc 30	70 **Ga** Gallium 31	73 **Ge** Germanium 32	75 **As** Arsenic 33	79 **Se** Selenium 34	80 **Br** Bromine 35	84 **Kr** Krypton 36
5	85.5 **Rb** Rubidium 37	88 **Sr** Strontium 38	89 **Y** Yttrium 39	91 **Zr** Zirconium 40	93 **Nb** Niobium 41	96 **Mo** Molybdenum 42	98 **Tc** Technetium 43	101 **Ru** Ruthenium 44	103 **Rh** Rhodium 45	106 **Pd** Palladium 46	108 **Ag** Silver 47	112 **Cd** Cadmium 48	115 **In** Indium 49	119 **Sn** Tin 50	122 **Sb** Antimony 51	128 **Te** Tellurium 52	127 **I** Iodine 53	131 **Xe** Xeron 54
6	133 **Cs** Caesium 55	137 **Ba** Barium 56	139 **La** Lanthanium 57	178.5 **Hf** Hafnium 72	181 **Ta** Tantalum 73	184 **W** Tungsten 74	186 **Re** Rhenium 75	190 **Os** Osmium 76	192 **Ir** Iridium 77	195 **Pt** Platinum 78	197 **Au** Gold 79	210 **Hg** Mercury 80	204 **Tl** Thallium 81	207 **Pb** Lead 82	209 **Bi** Bismuth 83	210 **Po** Polonium 84	210 **At** Astatine 85	222 **Rn** Radon 86
7	223 **Fr** Francium 87	226 **Ra** Radium 88	227 **Ac** Actinium 89	**Db** Dubnium 104	**Jl** Juliotum 105	**Rf** Rutherfordium 106	**Bh** Bohrium 107	**Hn** Hahnium 108	**Mt** Meitnerium 109									

KEY:

Atomic mass
Symbol
Name
Atomic Number

139 **La** Lanthanium 57	140 **Ce** Cerium 58	141 **Pr** Praseodymium 59	144 **Nd** Neodymium 60	147 **Pm** Promethium 61	150 **Sm** Samarium 62	152 **Eu** Europium 63	157 **Gd** Gadolinium 64	159 **Tb** Terbium 65	162.5 **Dy** Dysprosium 66	165 **Ho** Holmium 67	167 **Er** Erbium 68	169 **Tm** Thulium 69	173 **Yb** Ytterbium 70	175 **Lu** Lutetium 71
227 **Ac** Actinium 89	232 **Th** Thorium 90	231 **Pa** Procactinium 91	238 **U** Uranium 92	237 **Np** Neptunium 93	242 **Pu** Plutonium 94	243 **Am** Americium 95	247 **Cm** Curium 96	247 **Bk** Berkelium 97	251 **Cf** Californium 98	254 **Es** Einsteinium 99	253 **Fm** Fermium 100	256 **Md** Mendeleevium 101	254 **No** Nobelium 102	257 **Lw** Lawrencium 103

The vertical columns in the periodic table are called **groups**. **The elements in a group have similar properties**.

In OCR, Roman numerals are used for group numbers e.g. halogens – group VII.

The horizontal rows of elements are called **periods**.

The **main block** of elements are shaded in **Fig. 3.1**. The elements between the two parts of the main block are the **transition metals**.

The bold stepped line on the table divide **metals** on the left hand side from **non-metals** on the right.

Development of the periodic table

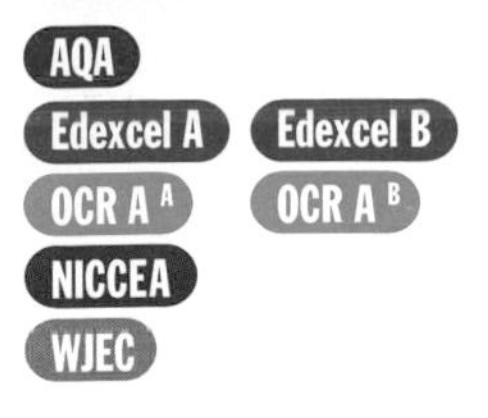

In the early 19th century many new elements were being discovered and chemists were looking for similarities between these new elements and existing elements.

Döbereiner (1829) suggested that elements could be grouped in threes **(triads)**. Each member of the triad had similar properties.

E.g. lithium, sodium, potassium
chlorine, bromine, iodine

Newlands (1863) arranged the elements in order of **increasing relative atomic mass**. He noticed that there was some similarity between every eighth element.

Li	Be	B	C	N	O	F	
Na	Mg	Al	Si	P	S	Cl,	etc

These were called **Newlands' Octaves**. Unfortunately the pattern broke down with the heavier elements and because he left no gaps for undiscovered elements. His work did not receive much support at the time.

Meyer (1869) looked at the relationship between **atomic mass** and the **density of an element**. He then plotted a graph of atomic volume (relative atomic mass in g divided by density) against the relative atomic mass for each element. The curve he obtained showed a series of peaks and troughs. **Elements with similar properties were in similar places on the graph.**

It was Mendeleev's foresight which was the major step forward.

Mendeleev arranged the elements in order of increasing relative **atomic mass**, but took into account the patterns of behaviour of the elements. He found it was necessary to leave gaps in the table and said that these were for elements not known at that time. His table enabled him to predict the properties of the undiscovered elements. His work was proved correct by the accurate prediction of the properties of gallium and germanium. The periodic table we use today closely resembles the table drawn up by Mendeleev.

A modification of the periodic table was made following the work of **Rutherford** and **Moseley**. It was realised that the elements should be arranged in order of atomic number, i.e. the number of protons in the nucleus. In the modern periodic table the elements are arranged in order of increasing atomic number with elements with similar properties in the same column.

Relationship between electron arrangement and position in the periodic table

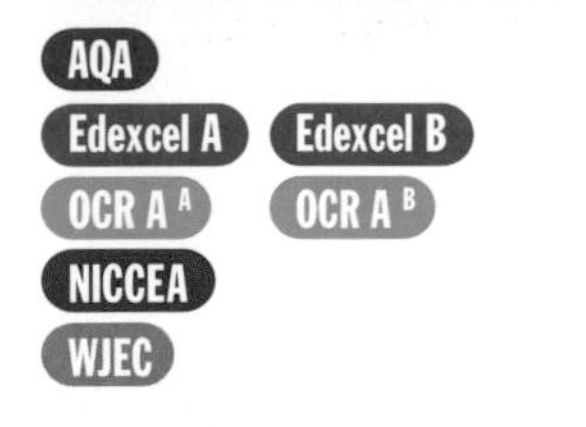

KEY POINT

For any element in the main block of the periodic table, it is easy to work out the electron arrangement in the atoms.
- **The number of energy levels or shells is the same as the period in which the element is placed.**
- **The number of electrons in the outer energy level is the same as the group number (except for elements in group 0 which have 8 electrons, apart from helium which has two electrons).**

For GCSE, you should be able to work out the electron arrangements of the first 20 elements.

Strontium is in period 5 and group 2. This means there are five energy levels used and two electrons in the outer energy level.

If you look up the electron arrangement of strontium, it is 2, 8, 18, 8, 2

Table 3.1 shows the arrangement of electrons in atoms of alkali metals (group 1).

Element	Atomic number	Electron arrangement
Li	3	2,1
Na	11	2,8,1
K	19	2,8,8,1
Rb	37	2,8,18,8,1
Cs	55	2,8,18,18,8,1

Note that, in each case, the outer energy level contains just one electron. When an element reacts, it attempts to obtain a full outer energy level.

Group I elements will lose one electron when they react and will form a positive ion.

$Na \rightarrow Na^{+} + e^{-}$

We can explain the order of reactivity within the group. The electrons are held in position by the electrostatic attraction of the positive nucleus. **This means that the closer the electron is to the nucleus, the harder it will be to remove it**.

As we go down the group, the outer electron gets further away from the nucleus and **so becomes easier to take away**. This means as we go down the group, the reactivity should increase.

Table 3.2 shows the arrangement of electrons in atoms of halogens (group 7).

Element	Atomic number	Electron arrangement
F	9	2,7
Cl	17	2,8,7
Br	35	2,8,18,7
I	53	2.8.18.18.7

Note that each member of the halogens (group 7) has seven electrons in the outer energy level. This is just one electron short of the full energy level. When halogen elements react, they gain an electron to complete that outer energy level. This will form a negative ion.

The fact that reactivity increases down group 1 but decreases down group 7 frequently leads to mistakes by students.

$Cl + e^- \rightarrow Cl^-$

As an electron is being gained in the reaction, the most reactive member of the family will be the one where the extra electron is closest to the nucleus,i.e. fluorine.

The reactivity decreases down the group.

Properties and reactions of alkali metals

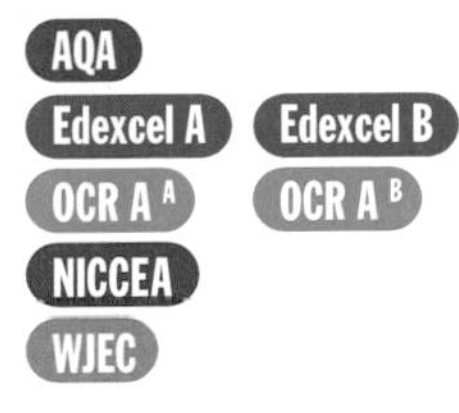

The alkali metals are a family of **very reactive metals**. The most common members of the family are lithium ,sodium and potassium. Some of the properties of these elements are shown in **Table 3.3** below.

Element	Symbol	Appearance	Melting point in °C	Density In g/cm³
Lithium	Li	Soft grey metal	181	0.54
Sodium	Na	Soft light grey metal	98	0.97
Potassium	K	Very soft blue/grey metal	63	0.86

These metals have to be stored in oil to exclude air and water. They do not behave much like metals, at first sight, but when **freshly cut they all have a typical shiny metallic surface**.

They are also **very good conductors of electricity**. Note, however, that they have **melting points and densities that are low** compared with other metals.

In the periodic table the alkali metals are in group 1.

Reaction of alkali metals with water

When a small piece of an alkali metal is put into a trough of water, the metal reacts immediately, floating on the surface of the water and evolving **hydrogen**.

With sodium and potassium, the heat evolved from the reaction is sufficient to melt the metal.

The hydrogen evolved by the reaction of potassium with cold water is usually ignited and burns with a pink flame.

Sodium reacts more quickly than lithium, and potassium reacts more quickly than sodium.

In each case the solution remaining at the end of the reaction is an alkali.

Examiners expect you to have seen this reaction. It can be viewed on many videos and CD-ROMs. Frequently the oxide is given as the product rather than the hydroxide.

$2Li(s) + 2H_2O(l) \rightarrow 2LiOH(aq) + H_2(g)$

lithium + water → lithium hydroxide + hydrogen

$2Na(s) + 2H_2O(l) \rightarrow 2NaOH(aq) + H_2(g)$

sodium + water → sodium hydroxide + hydrogen

$2K(s) + 2H_2O(l) \rightarrow 2KOH(aq) + H_2(g)$

potassium + water → potassium hydroxide + hydrogen

N.B. These three equations are basically the same and, if the alkali metal is represented by M, these equations can be represented by:

$\mathbf{2M(s) + 2H_2O(l) \rightarrow 2MOH(aq) + H_2(g)}$

Reaction of alkali metals with oxygen

When heated in air or oxygen, the alkali metals burn to form white solid **oxides**. The colour of the flame is characteristic of the metal:

lithium — red

sodium — orange

potassium — lilac

E.g. $4Li(s) + O_2(g) \rightarrow 2Li_2O(s)$

lithium + oxygen → lithium oxide

or $4M(s) + O_2(g) \rightarrow 2M_2O(s)$

The **alkali metal oxides** all dissolve in water to form **alkali solutions**.

E.g. $Li_2O(s) + H_2O(l) \rightarrow 2LiOH(aq)$

lithium oxide + water → lithium hydroxide

or $M_2O(s) + H_2O(l) \rightarrow 2MOH(aq)$

Reaction of alkali metals with chlorine

When a piece of burning alkali metal is lowered into a gas jar of chlorine, the metal continues to burn forming a white smoke of the metal **chloride**.

E.g. $2K(s) + Cl_2(g) \rightarrow 2KCl(s)$

potassium + chlorine → potassium chloride

or $2M(s) + Cl_2(g) \rightarrow 2MCl(s)$

It is because of these similar reactions that these metals are put in the same family. In each reaction the order of reactivity is the same, i.e. **lithium is the least reactive and potassium is the most reactive**.

There are three more members of this family: rubidium (Rb), caesium (Cs) and francium (Fr). They are all more reactive than potassium.

Properties and reactions of halogens

The halogens are a family of non-metals.

KEY POINT

In the halogen family, the different elements have different appearances but they are in the same family on the basis of their similar chemical properties.

Table 3.4 compares the appearances of four of these elements.

Candidates frequently spell fluorine as flourine.

Element	Symbol	Appearance at room temperature
Fluorine	F	Pale yellow gas
Chlorine	Cl	Yellow/green gas
Bromine	Br	Red/brown volatile liquid
Iodine	I	Dark grey crystalline solid

There is another member of the family called astatine (At). It is radioactive and a very rare element.

Fluorine is a very reactive gas and is too reactive to handle in normal laboratory conditions.

In the periodic table the halogens are in group 7.

Solubility of halogens in water

None of the halogens is very soluble in water. Chlorine is the most soluble. Iodine does not dissolve much in cold water and only dissolves slightly in hot water.

Halogens contain molecules with covalent bonding. They dissolve better in organic solvents e.g. hexane.

Chlorine solution (sometimes called chlorine water) is very pale green. It turns Universal Indicator red, showing the solution is **acidic**. The colour of the indicator is quickly bleached.

Bromine solution (bromine water) is orange. It is very weakly acidic and also acts as a bleach.

Iodine solution is very weakly acidic and is also a slight bleach. The low solubility of halogens in water (a polar solvent) is expected because halogens are composed of molecules.

Solubility of halogens in hexane (a non-polar solvent)

The halogens dissolve readily in hexane to give solutions of characteristic colour:

chlorine – colourless

bromine – orange

iodine – purple

Reactions of halogens with iron

The halogens react with **metals** by direct combination to form **salts**. The name **'halogen' means salt producer. Chlorine** forms **chlorides, bromine** forms **bromides and iodine** forms **iodides.**

If **chlorine** gas is passed over heated **iron** wire, an exothermic reaction takes place forming **iron(III) chloride**, which forms as a brown solid on cooling.

Fig. 3.2 shows a suitable apparatus for preparing anhydrous iron(III) chloride crystals.

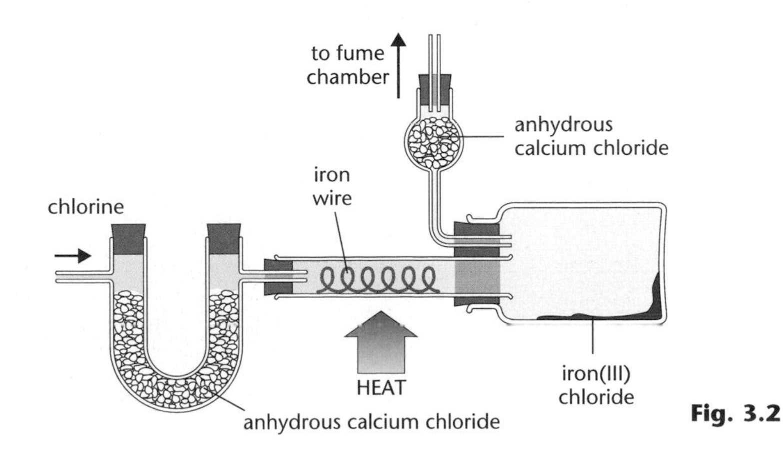

Fig. 3.2

$2Fe(s) + 3Cl_2(g) \rightarrow 2FeCl_3(s)$

iron + chlorine → iron(III) chloride

Bromine vapour also reacts with hot iron wire to form iron(III) bromide. When iodine crystals are heated, they turn to a purple vapour. This vapour reacts with hot iron wire to produce iron(II) iodide.

Order of reactivity of the halogens

From their chemical reactions the relative reactivities of the halogens are:

The reactivity of halogens decreases down the group.

fluorine **most reactive**

chlorine

bromine

iodine **least reactive**

Displacement reactions of the halogens

KEY POINT **A more reactive halogen will displace a less reactive halogen from one of its compounds.**

For example, when **chlorine** is bubbled into a solution of **potassium bromide**, the chlorine displaces the less reactive bromine. This means the colourless solution turns orange as the free bromine is formed.

$2KBr(aq) + Cl_2(g) \rightarrow 2KCl(aq) + Br_2(aq)$

potassium bromide + chlorine → potassium chloride + bromine

No reaction would take place if iodine solution were added to potassium bromide solution because iodine is less reactive than bromine.

Properties and uses of the noble gases

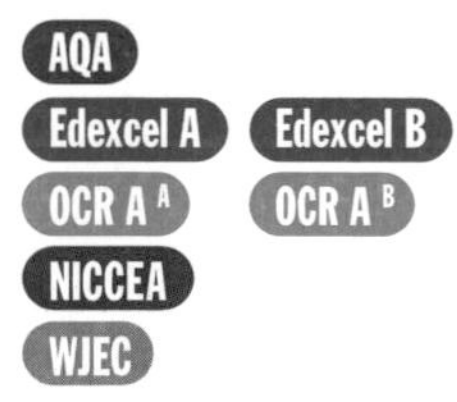

The noble gases are a family of unreactive gases in group 0 of the periodic table. They were not known when Mendeleev devised the first periodic table.

The reason they were not discovered earlier is they are very unreactive. Until about 40 years ago it was believed that they never reacted. We now know that they form some compounds, e.g. xenon tetrafluoride, XeF_4.

Table 3.5 gives some information about noble gases.

Element	Symbol	Boiling point (°C)		Density (g/dm³)	
Helium	He	–270	boiling point increases ↓	0.17	boiling point increases ↓
Neon	Ne	–249		0.84	
Argon	Ar	–189		1.66	
Krypton	Kr	–157		3.46	
Xenon	Xe	–112		5.46	
Radon	Rn	–71		8.9	

Table 3.6 gives some uses of noble gases.

Most of these uses rely upon the unreactivity of noble gases.

Noble gas	Use
Helium	Balloons and airships – less dense than air and not flammable
Neon	Filling advertising tubes
Argon	Filling electric light bulbs – inert atmosphere for welding
Krypton and xenon	Lighthouse and projector bulbs. Lasers
Radon	Killing cancerous tumours

Properties and uses of transition metals

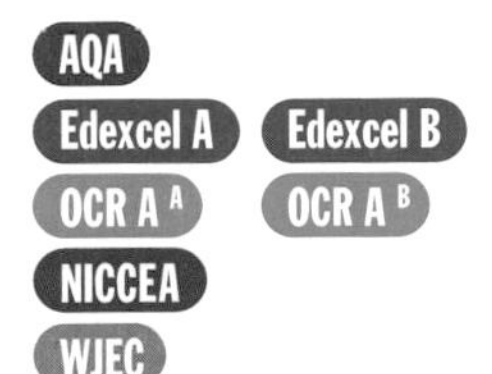

The transition metals are in a block of metals between groups 2 and 3 in the periodic table.

Iron, nickel and manganese are examples. These metals have a number of features in common including:

Do not confuse the transition metal manganese with magnesium, a metal in group 2.

- **higher melting points**, **boiling points** and **densities** than group 1 metals
- usually **shiny** appearance
- **good conductors of heat and electricity**
- some have **strong magnetic properties**
- often **form more than one positive ion**. For example, iron forms iron(II) ions, Fe^{2+}, and iron(III) ions, Fe^{3+}.
- **compounds are often coloured**. For example, iron(II) sulphate is pale green and iron(III) sulphate is yellow-brown
- transition metals and transition metal compounds are often good **catalysts**. For example, iron is used as the catalyst in the Haber process to produce ammonia.

Transition metal oxides are used to make coloured glazes for pottery.

Transition metals have a wide range of uses, either as pure metals or in mixtures of metals called **alloys**.

Alloys have better properties for most uses than pure metals.

Pure metals are used for electrical conductors as pure metals conduct electricity better.

Steel is an alloy of **iron with a small percentage of carbon**. It is used for making car bodies, ships and bridges. Steel rods are also used to reinforce concrete. **Stainless steel** contains other transition metals such as **nickel** and **chromium**. It is more resistant to corrosion than ordinary steel.

Brass is an alloy of **copper** and **zinc**. It is used for door handles, hinges and decorative ware.

Bronze is an alloy of **copper** and **tin**. It is used to make statues.

Gold and **silver** are used for jewellery, but again they are hardened by alloying with other metals.

Colours of transition metal hydroxides

Transition metal hydroxides are often precipitated when sodium hydroxide is added to a solution of a transition metal compound.

These metal hydroxides have characteristic colours

E.g. $CuSO_4 + 2NaOH \rightarrow Cu(OH)_2 + Na_2SO_4$

Copper(II) sulphate + sodium hydroxide →
Copper(II) hydroxide + sodium sulphate

Copper(II) hydroxide is a blue precipitate.

$FeSO_4 + 2NaOH \rightarrow Fe(OH)_2 + Na_2SO_4$

Iron(II) sulphate + sodium hydroxide →
Iron(II) hydroxide + sodium sulphate

Iron(II) hydroxide is a dirty green precipitate.

$Fe_2(SO_4)_3 + 6NaOH \rightarrow 2Fe(OH)_3 + 3Na_2SO_4$

Iron(III) sulphate + sodium hydroxide →
Iron(III) hydroxide + sodium sulphate

Iron(III) hydroxide is a red-brown precipitate.

PROGRESS CHECK

Here is a list of elements. Use your periodic table to answer these questions.
chlorine helium lithium magnesium titanium

1. Which element is in period 1?
2. Which element is in group 2?
3. Which element is an alkali metal?
4. Which element is a halogen?
5. Which element is a transition metal?
6. Which element is a noble gas?
7. Which elements have atoms containing two electrons in the outer energy level?
8. Which two of these elements are in the same period of the periodic table?
9. Sterling silver is an alloy used to make jewellry . Silver is mixed with copper. What are two reasons why sterling silver is better than pure silver for jewellery.

1. Helium; 2. Magnesium; 3. Lithium; 4. Chlorine; 5. Titanium; 6. Helium; 7. Helium and magnesium; 8. Magnesium and chlorine; 9. Sterling silver is cheaper than pure silver. It is harder than pure silver.

3.2 Chemical reactions

After studying this section you should be able to:

- ***classify reactions in various types e.g. neutralisation, oxidation, etc.***
- ***predict reactions using patterns in the properties of elements.***

Types of chemical reaction

AQA
Edexcel A Edexcel B
OCR A [A] OCR A [B]
NICCEA
WJEC

Table 3.7 contains examples of different types of reaction. In each case there is an example of how the reaction can be used to produce new materials.

Table 3.7 Types of chemical reaction

Type of reaction	Definition	Example of its use
Neutralisation	The reaction of an acid with a base, or alkali, to form a salt and water only	Sodium hydroxide reacts with hydrochloric acid to form sodium chloride and water only: $NaOH + HCl \rightarrow NaCl + H_2O$
Oxidation	A reaction which involves the addition of oxygen or the loss of hydrogen	Ethanol is oxidised by oxygen in the air to produce ethanoic acid. This reaction occurs when wine turns to vinegar: $C_2H_5OH + O_2 \rightarrow CH_3COOH + H_2O$
Reduction	A reaction that involves the addition of hydrogen or the loss of oxygen	Copper(II) oxide is heated in a stream of hydrogen gas: $CuO + H_2 \rightarrow Cu + H_2O$
Thermal decomposition	Decomposition is the splitting up of a compound. Thermal decomposition is the splitting up by heating	Calcium carbonate is decomposed by heating to produce calcium oxide: $CaCO_3 \rightarrow CaO + CO_2$
Precipitation	A solid is formed when two solutions are mixed. The solid is called a precipitate	Barium sulphate is precipitated when solutions of barium nitrate and sulphuric acid are mixed: $Ba(NO_3)_2 + H_2SO_4 \rightarrow BaSO_4 + 2HNO_3$
Combustion	A reaction with oxygen usually accompanied by a release of energy. Combustion reactions are examples of oxidation reactions	Burning carbon can produce carbon dioxide: $C + O_2 \rightarrow CO_2$

Many questions involve you in making predictions about reactions from the data given. it is important that you use this data fully.

Other types of reaction include:
addition (page 28),
polymerisation (page 28),
cracking (page 27),
exothermic (page 27),
endothermic (page 73),
displacement reactions (page 60)

Patterns in chemical properties can be used to predict reactions. Examples include:

1. displacement reactions of halogens (page 60).
2. patterns within groups in the periodic table (pages 57–61).
3. reactivity series of metals.

If a piece of iron is dipped into copper(II) sulphate solution a reaction takes place because iron is higher in the reactivity series than copper.

Brown copper is precipitated and the blue solution fades.

$$CuSO_4 + Fe \rightarrow FeSO_4 + Cu$$

Choose answers from this list:
decomposition neutralisation oxidation precipitation reduction

For each of the reactions 1–5, choose the best word to describe the type of reaction.

1. $2Pb(NO_3)_2 \rightarrow 2PbO + 4NO_2 + O_2$
2. $2NH_3 + H_2SO_4 \rightarrow (NH_4)_2SO_4$
3. $AgNO_3(aq) + NaCl(aq) \rightarrow AgCl(s) + NaNO_3(aq)$
4. $C_2H_4 + H_2 \rightarrow C_2H_6$
5. $2Mg + O_2 \rightarrow 2MgO$

1. Decomposition; 2. Neutralisation; 3. Precipitation; 4. Reduction; 5. Oxidation

3.3 Rates of reaction

After studying this section you should be able to:

- ***recall the conditions that can be altered to change the rate of a reaction***
- ***describe how an experiment can be used to demonstrate the effect of changing one of the conditions***
- ***explain, using ideas of particles why changing a condition alters the rate of reaction***
- ***describe and explain how enzymes can be used in industrial processes.***

Reactions at different rates

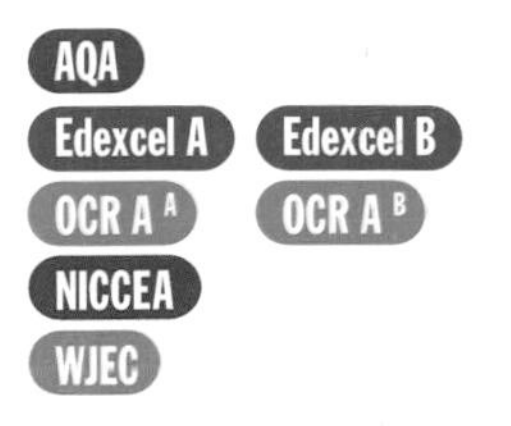

There are chemical reactions that take place very quickly and ones that take place very slowly.

When a lighted splint is placed in a mixture of hydrogen and air, an explosion takes place and a squeaky pop is heard. This reaction is over in a tiny fraction of a second. It is a very fast reaction.

Candidates often confuse rate and time. If a reaction takes longer, the rate decreases.

A limestone building reacts with acidic gases in the air. This reaction takes hundreds of years before the effects can be seen. This is a very slow reaction.

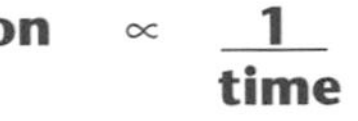

$$\textbf{Rate of reaction} \propto \frac{\textbf{1}}{\textbf{time}}$$

For practical reasons, reactions used in the laboratory for studying rate of reaction must not be too fast or too slow.

Having selected a suitable reaction, it is necessary to find a change that can be observed during the reaction. An estimate of the rate of reaction can be found from the time for a measurable change to take place.

Measuring the volume of gas at intervals

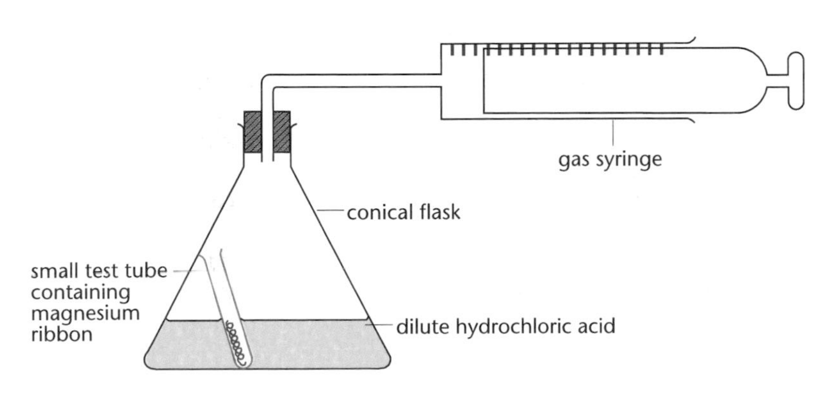

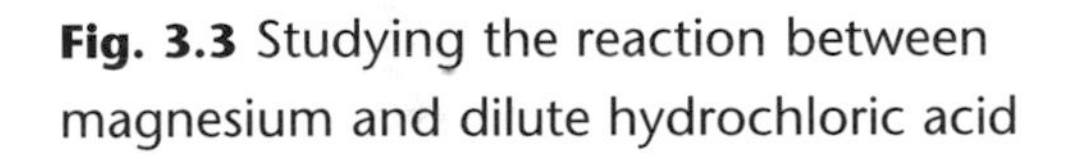

Fig. 3.3 Studying the reaction between magnesium and dilute hydrochloric acid

Some of the easiest reactions to study in the laboratory are those where a gas is evolved. The reaction can be followed by measuring the volume of gas evolved over a period of time using the apparatus in **Fig. 3.3**.

A good example is the reaction of magnesium with dilute hydrochloric acid:

$Mg(s) + 2HCl(aq) \rightarrow MgCl_2(aq) + H_2(g)$

magnesium + hydrochloric acid → magnesium chloride + hydrogen

It is important to keep the reactants separate whilst setting up the apparatus so that the starting time of the reaction can be measured accurately.

Fig. 3.4 shows a typical graph obtained for the reaction between dilute hydrochloric acid and magnesium.

N.B. If you are carrying out an experiment, not all of the points may lie on the curve. This is because of experimental error. You should draw the best line through, or close to, as many points as possible.

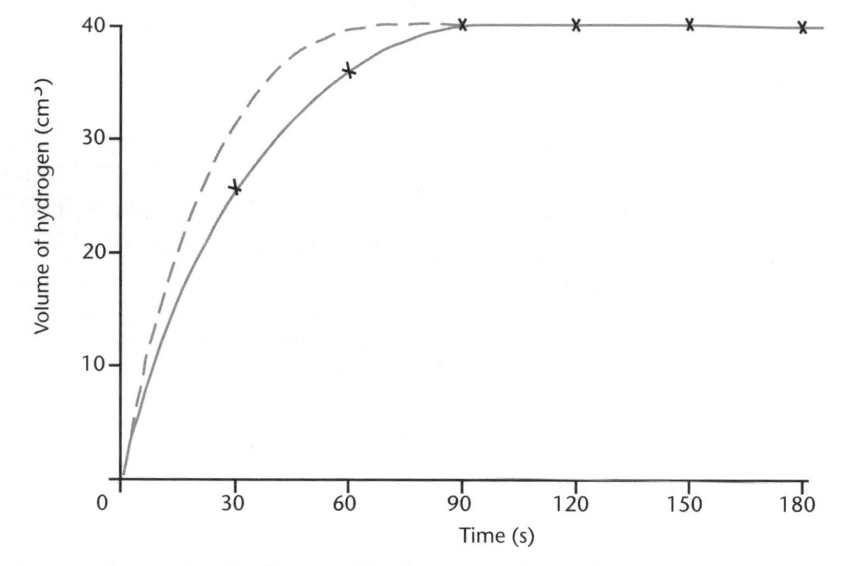

Fig. 3.4 A graph of volume of hydrogen collected at intervals

The dotted line shows the graph for a similar experiment using the same quantities of magnesium and hydrochloric acid, but with conditions changed so that the reaction is slightly faster. The **rate of the reaction is greatest** when the graph is **steepest**, i.e. at the start of the reaction. The reaction is finished when the graph becomes **horizontal**, i.e. there is no further increase in the volume of hydrogen.

It is often possible to follow the course of similar reactions by measuring the loss of mass during the reaction due to escape of gas.

For example, if calcium carbonate and hydrochloric acid are used, the loss of mass of calcium carbonate is significant. However, with magnesium and hydrochloric acid, the loss of mass is very small.

There are ICT opportunities here. The volume of carbon dioxide collected or the mass loss can be found by using a computer.

Other suitable changes that can be measured include:

- **colour changes**
- **formation of a precipitate**
- **time taken for a given mass of solid to react**
- **pH changes**
- **temperature changes.**

Factors affecting rate of reaction

AQA
Edexcel A Edexcel B
OCR A [A] OCR A [B]
NICCEA
WJEC

Table 3.8 compares some of the factors that affect the rate of chemical reactions.

Factor	Reactions affected	Change made in conditions	Effect on rate of reactions
Temperature	All	Increase 10 °C Decrease 10 °C	Approx. doubles rate Approx. halves rate
Concentration	All	Increase in concentration of one of the reactants	Increases the rate of reaction
Pressure	Reactions involving mixtures of gases	Increase the pressure	Greatly increases the rate of reaction
Light	Wide variety of reactions including reactions with mixtures of gases including chlorine or bromine	Reaction in sunlight or uv light	Greatly increases the rate of reaction
Particle size	Reactions involving solids and liquids, solids and gases, or mixtures of solids	Using one or more solids in a powdered form	Greatly increases the rate of reaction
Using a catalyst	Adding a substance to a reaction mixture	A specific substance which speeds up the reaction without being used up	Increases rate of reaction

Explaining different rates using particle model

AQA
Edexcel A Edexcel B
OCR A [A] OCR A [B]
NICCEA
WJEC

Before looking how the rate of reaction can be changed by altering one factor in **Table 3.8**, we must first look at what happens to particles when a reaction takes place.

Particles in solids, liquids and gases are **moving**. This movement is much greater in gases than in liquids and in liquids more than in solids.

In a reaction mixture the particles of the reactants **collide**. Not every collision leads to reaction. Before a reaction occurs, the particles must have a **sufficient amount of energy**. This is called the **activation energy**. If a collision between particles can produce sufficient energy, i.e. if they collide fast enough and in the right direction, a reaction will take place. Not all collisions will result in a reaction.

A reaction is speeded up if the number of collisions is increased.

Fig. 3.5 shows an energy level diagram of a typical reaction.

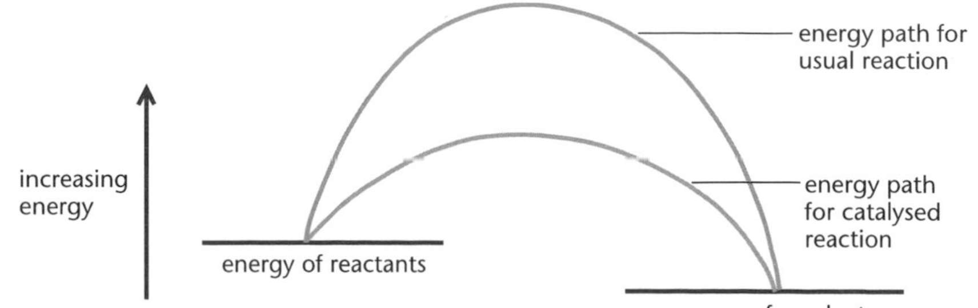

Fig. 3.5

High level answers explaining changes in rates of reaction should include understanding of particle movement.

Increasing the concentration

If concentration is increased, there are more collisions between particles and so there are more collisions leading to reaction and the reaction is faster.

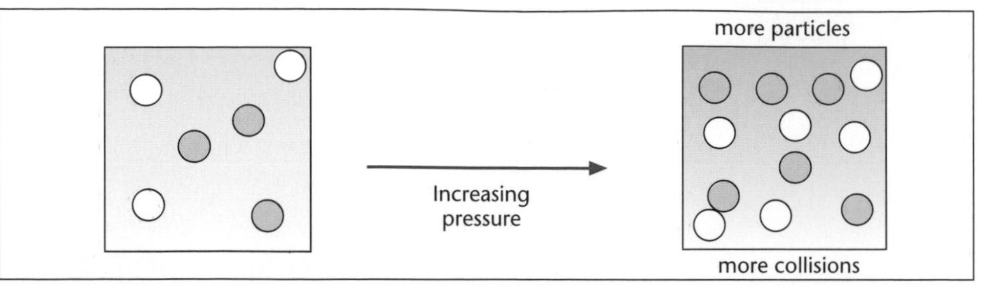

Fig. 3.6 Effect of pressure on reaction rate

Increasing the pressure can be explained in the same way, because increasing the pressure of a mixture of gases increases the concentration by forcing the particles closer together.

Increasing the temperature

Increasing the temperature makes the particles move **faster**. This leads to **more collisions**. Also, the particles have more kinetic energy, so more collisions will lead to reaction. Using sunlight or uv light has the same effect as increasing temperature.

Using smaller pieces of solid

When one of the reactants is a solid, the reaction must take place on the surface of the solid. By breaking the solid into smaller pieces, the **surface area** is **increased**, giving a greater area for collisions to take place and so causing an increase in the rate of reaction.

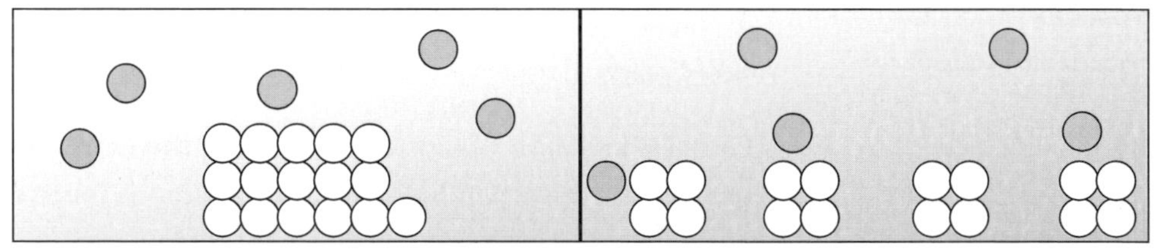

Smaller pieces have a large surface area. More collisions so faster reaction

Fig. 3.7 Effect of surface area on reaction rate

Using a catalyst

A catalyst is a substance that alters the rate of reaction but remains chemically unchanged at the end of the reaction.

Catalysts usually speed up reactions. A catalyst which slows down a reaction is called a negative catalyst or **inhibitor**.

Manganese(IV) oxide catalyses the decomposition of hydrogen peroxide into water and oxygen:

$$2H_2O_2(aq) \rightarrow 2H_2O(l) + O_2(g)$$

Catalysts are often transition metals or transition metal compounds.
The catalyst provides a **surface** where the reaction can take place.

Using a catalyst lowers the activation energy for the reaction. More collisions have sufficient energy for reactions to take place.

OCR A does not use the term activation energy.

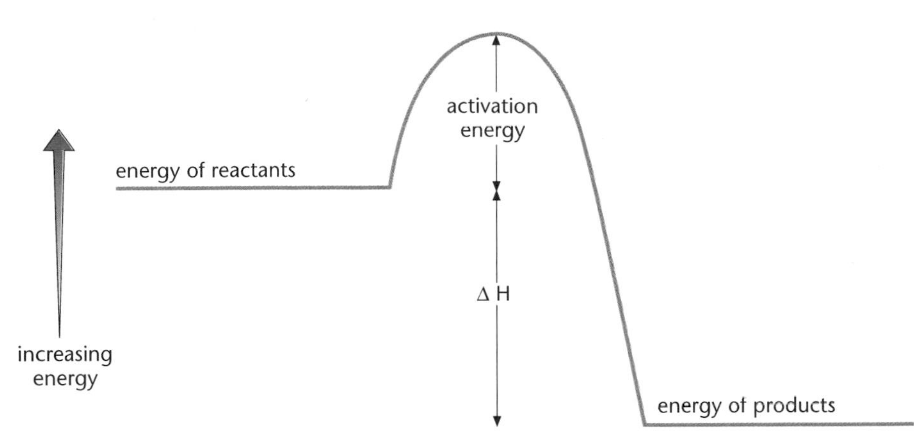

Fig. 3.8 Activation energy

Enzymes

AQA
Edexcel A Edexcel B
OCR A[A] OCR A[B]
NICCEA
WJEC

Enzymes are **biological catalysts**. Hydrogen peroxide is decomposed into water and oxygen by an enzyme in fruits and vegetables. Catalase is a protein. Unlike chemical catalysts, such as manganese(IV) oxide, catalase works only under particular conditions. It works best at 37 °C. At higher temperatures the protein structure is permanently changed **(denatured)**: it no longer decomposes hydrogen peroxide.

Successful enzyme processes

- **Stabilise the enzyme, so it works for a long time**
- **Trap the catalyst**
- **Are continuous**

Enzymes are used in many industrial processes:

- fermentation of solutions of starch and sugar using enzymes in yeast to produce beer and wine
- making cheese and yogurt by the action of enzymes on milk
- enzymes (proteases and lipases) in washing powders break down protein stains in cold or warm water
- soft-centred chocolates are made by injecting hard-centred chocolates with the enzyme invertase
- isomerase is used to turn glucose syrup into fructose syrup. This is sweeter and can be used in smaller quantities in slimming products.

PROGRESS CHECK

Use ideas of rate of reaction to explain each of the following observations:

1. Mixtures of coal dust and air in coal mines can explode, but lumps of coal are difficult to set alight.
2. Milk takes longer to sour when kept in a refrigerator than when on the doorstep.
3. Vegetables cook faster in a pressure cooker.
4. Mixtures of methane and chlorine do not react in the dark but react in sunlight.
5. Some adhesives are sold in two tubes. The contents of the two tubes have to be mixed before the glue sets.
6. Chips fry faster in oil than potatoes cook in boiling water.

1. Coal dust has a very large surface area. The reaction is speeded up; 2. Lower temperature slows down the rate of souring; 3. In a pressure cooker, increasing the pressure lowers the boiling point of water and speeds up the reaction; 4. Light provides initial energy to speed up the reaction; 5. One tube contains a catalyst. Mixing the two tubes speeds up the setting; 6. Oil is at a much higher temperature than boiling water so reactions are faster.

3.4 Reversible reactions

LEARNING SUMMARY

After studying this section you should be able to:

- ***use given data to explain how yields can be maximised in industrial processes involving reversible reactions.***

Getting a maximum yield

AQA
Edexcel A Edexcel B
OCR A A OCR A B
NICCEA
WJEC

In a non-reversible reaction, it is possible to get 100% conversion of the reactants to products.

For example, in the combustion of 24 g of magnesium in excess oxygen, 40 g of magnesium oxide can be formed.

$2Mg + O_2 \rightarrow 2MgO$

In practice, however, complete conversion to products (called 100% yield) is rarely obtained. This is because some reactants and products are lost or other side-reactions take place.

If a reaction is **reversible**, 100% conversion of reactants to products is much more difficult. The amount converted (called **yield**) depends upon conditions.

The reaction of iron with steam is represented by the equation:

$3Fe(s) + 4H_2O(g) \rightleftharpoons Fe_3O_4(s) + 4H_2(g)$

If iron and steam is heated is a closed container, **equilibrium** is set up.

This means that the concentrations of the two reactants and the two products remain unchanged provided conditions are constant.

Apparently, the reaction has stopped. This is not the case, however.

The forward and reverse reactions are still taking place, but they are taking place at the same rate.

No more products will be obtained if the mixture is left, and 100% conversion is impossible.

It is possible to get greater conversion of reactants to products if reactions do not take place in closed containers.

If steam is passed over heated iron in such a way that the hydrogen produced escapes, more steam will react with the iron until all of the iron has reacted.

Contact process

AQA
Edexcel A Edexcel B
OCR A A OCR A B
NICCEA
WJEC

Sulphuric acid is made in industry in a three-stage process.

Stage 1: Burning sulphur or heating sulphide minerals in air:

$$S + O_2 \rightarrow SO_2$$

Stage 2: Reacting sulphur dioxide with oxygen to produce sulphur trioxide.

$$2SO_2 + O_2 \rightleftharpoons 2SO_3$$

Stage 3: Absorbing sulphur trioxide to form sulphuric acid:

$$SO_3 + H_2O \rightarrow H_2SO_4$$

You will not normally be required to recall the details of this process. You may be expected to comment on it when the information is given to you.

[In practice, this is done in two stages – the sulphur trioxide is dissolved in concentrated sulphuric acid and then diluted with the required amount of water to make concentrated sulphuric acid.]

Only stage 2 is reversible. Controlling this reaction is the secret to getting the maximum yield in the whole process.

Remember, a catalyst does not produce more. It produces the same amount but more quickly.

To get the best yield, low temperatures are desirable. However, low temperatures will slow reactions down. There has to be a **compromise** between getting a good yield and getting the yield quickly. A vanadium(V) oxide catalyst can be used to speed up the reaction.

1. What is the name of the industrial process used to manufacture ammonia?
2. What is the catalyst in the process producing ammonia?
3. What is the name of the industrial process used to make sulphuric acid?
4. What are the raw materials in the industrial process producing sulphuric acid?
5. Which two gases react together in the catalyst chamber to produce sulphur trioxide?
6. What is the catalyst in the process producing sulphuric acid?
7. Why are conditions important in the catalyst chamber?

1. Haber process; 2. Iron; 3. Contact process; 4. Sulphur, air and water; 5. Sulphur dioxide and oxygen; 6. Vanadium(V) oxide; 7. The conditions in the catalyst chamber determine the total yield in the process.

3.5 Energy transfer in reactions

LEARNING SUMMARY

After studying this section you should be able to:

- *recall that during exothermic reactions energy is lost to the surroundings during endothermic reactions and where energy is taken in from the surroundings*
- *recall that energy is required to break chemical bonds and energy is released when bonds are formed*
- *draw energy level diagrams for exothermic and endothermic reactions*
- *use bond energy data to calculate energy changes in reactions.*

Endothermic and exothermic reactions

AQA
Edexcel A Edexcel B
OCR A [A] OCR A [B]
NICCEA
WJEC

There are many examples where energy is either **released** or **taken** in during a chemical reaction.

Exothermic reactions

The burning of carbon in oxygen releases energy. Such a reaction is called an **exothermic reaction**.

The **quantity of energy** released depends upon the **mass** of carbon burned:

$C + O_2 \rightarrow CO_2$

carbon + oxygen → carbon dioxide ($\Delta H = -393.5$ kJ)

This information tells a chemist that burning 12 g of carbon in oxygen produces 393.5 kJ.

ΔH is called the **enthalpy** (or heat) **of reaction**. A **negative** value is used to show energy that is released.

The process is summarised in **Fig. 3.9**.

If two solutions are mixed at room temperature and the temperature rises, an exothermic reaction has taken place.

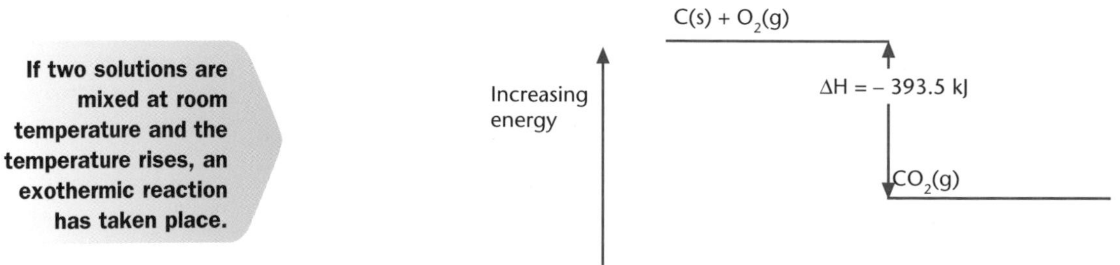

Fig. 3.9 Energy diagram for the complete combustion of carbon

Endothermic reactions

There are some reactions where energy is absorbed from the surroundings during the reaction and the temperature falls. These are called **endothermic** reactions.

For example, the formation of hydrogen iodide from hydrogen and iodine absorbs energy from the surroundings:

$H_2(g) + I_2(g) \rightleftharpoons 2HI(g)$

hydrogen + iodine $\rightleftharpoons$ hydrogen iodide ($\Delta H = +52$ kJ)

ΔH is **positive**, in this case, because the reaction is endothermic. This is summarised in **Fig. 3.10**.

Most reactions are exothermic. At one time scientists thought that only exothermic reactions could take place because they could not find any endothermic ones.

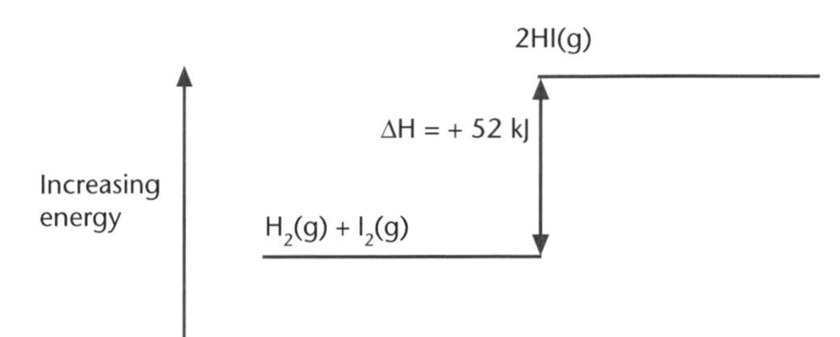

Fig. 3.10 Energy diagram for the formation of hydrogen iodide

Bond making and bond breaking

AQA
Edexcel A Edexcel B
OCR A [A] OCR A [B]
NICCEA
WJEC

During a chemical reaction there are changes in bonding taking place. Whether a particular reaction is exothermic or endothermic depends upon the energy required to break bonds and the energy released when bonds form.

If a reaction is an exothermic reaction:
energy released when bonds form > energy required to break bonds.
The surplus energy raises the temperature of the surroundings.
If a reaction is an endothermic reaction:
energy released when bonds form < energy required to break bonds.
The extra energy needed is taken from the surroundings so the temperature of the surroundings falls.

E.g.

Hydrogen and chlorine react together to form hydrogen chloride:

$H_2(g) + Cl_2(g) \rightarrow 2HCl(g)$

Using relative atomic masses, 2 g of hydrogen react with 71 g of chlorine to form 73 g of hydrogen chloride.

This can be represented by:

H–H Cl–Cl $\rightarrow$ H–Cl H–Cl

We can find out the energy required to break 2 g of hydrogen molecules and 71 g of chlorine molecules by looking up data in a data book. Also, we can find the energy released when 73 g of hydrogen chloride is formed:

Energy required to break H–H bonds in 2 g of hydrogen molecules = +436 kJ

Energy required to break Cl–Cl bonds in 71 g of hydrogen molecules = +242 kJ

From a data book, we can find out the energy released when 36.5 g of hydrogen chloride is formed:

When bonds are formed, the value is positive. When bonds are broken the value is negative.

Energy released when forming H–Cl bonds in 36.5g of hydrogen molecules = –431 kJ

The energy released when forming H–Cl bonds in 71 g (i.e. 2 × 36.5g) of hydrogen molecules = 2 x (+431) = –862 kJ

Energy change = (+436) + (+242) – 862 = –184 kJ

The negative value tells us that the reaction is exothermic.

PROGRESS CHECK

1. When sodium carbonate and calcium chloride solutions are mixed, the temperature of the solution dropped by 4°C. Is the reaction exothermic or endothermic?
2. Are combustion reactions exothermic or endothermic?
3. Is energy needed or given out when bonds are broken?
4. Is energy needed or given out when bonds are formed?
5. Are ΔH values for an exothermic reaction positive or negative?
6. Are ΔH values for an endothermic reaction positive or negative?
7. Suggest why the following reaction will be endothermic:
   $N_2 + O_2 \rightarrow 2NO$

1. Endothermic; 2. Exothermic; 3. Needed; 4. Given out; 5. Negative; 6. Positive;
7. Strong N = N and O = O bonds have to be broken and only two NO bonds formed.

Sample GCSE questions

1. A section of the periodic table is shown below.

${}^{1}_{1}$H							${}^{4}_{2}$He
${}^{7}_{3}$Li	${}^{9}_{4}$Be	${}^{11}_{5}$B	${}^{12}_{6}$C	${}^{14}_{7}$N	${}^{16}_{8}$O	${}^{19}_{9}$F	${}^{20}_{10}$Ne
${}^{21}_{11}$Na	${}^{24}_{12}$Mg	${}^{27}_{13}$Al	${}^{28}_{14}$Si	${}^{31}_{15}$P	${}^{32}_{16}$S	${}^{35.5}_{17}$Cl	${}^{40}_{18}$Ar
${}^{39}_{19}$K	${}^{40}_{20}$Ca					${}^{80}_{35}$Br	${}^{84}_{36}$Kr

(a) Using only the elements shown in this section of the table, write down the symbol for:

(i) a non-metal which is solid at room temperature. **[1]**

C ✓

B, Si , P or S would be other correct answers to the question. You must give the symbol and not the name.

(ii) a liquid element at room temperature. **[1]**

Br ✓

(iii) an element which forms an ion with two positive charges. **[1]**

Mg ✓

Ca would be a correct alternative answer. The examiner would also accept Be although this does not readily form ions

(iv) an element which is a gas containing single atoms. **[1]**

He ✓

Ne ,Ar or Kr would be correct alternatives.

(v) an element which forms an oxide having a formula of the type X_2O_3. **[1]**

Al ✓

B would be an alternative.

(b) (i) Name one substance with which all the elements with atomic number 3, 11 and 19 will react **[1]**

Oxygen ✓

Water or chlorine would be alternatives.

(ii) Explain why the elements 3,11 and 19 have similar chemical reactions.

Use your knowledge of atomic structure in your answer **[3]**

Atoms of all three elements have a single electron in the outer energy level ✓. Li 2,1; Na 2,8,1; K 2,8,8,1 ✓ This outer electron is lost each time ✓.

It is not enough to write that the elements are in the same group of the periodic table as that would not be using the ideas of atomic structure.

N.B. You are not asked to explain differences.

Sample GCSE questions

2. The order of reactivity of the halogens is:

Fluorine — most reactive
Chlorine
Bromine
Iodine — least reactive

The table below summarises the results of reactions when halogens are added to solutions of potassium halides. The table is unfinished.

Halogen added	Solutions of		
	Potassium chloride	**Potassium bromide**	**Potassium iodide**
Bromine	✘	✘	✓
Chlorine	✘		
Iodine			✘

(a) For each of these reactions, write **yes** if the reaction takes place or **no** if it does not.

(i) potassium bromide and chlorine **[1]**

Yes ✓

(ii) potassium iodide and chlorine **[1]**

Yes ✓

(iii) potassium chloride and iodine **[1]**

No ✓

(iv) potassium bromide and iodine **[1]**

No ✓

Answering these questions involves using the pattern of reactivity of the halogens given at the start of the question.

(b) What type of reaction is taking place when potassium iodide reacts with bromine? **[1]**

Displacement reaction ✓

(c) Write a balanced symbol equation for the reaction of potassium iodide and bromine. **[3]**

$2KI + Br_2 \rightarrow 2KBr + I_2$ ✓✓✓

The three marks here are for:
1. *the correct formulae of the reactants*
2. *the correct formulae of the products*
3. *balancing correctly.*

Exam practice questions

1. This question is about the elements in group 2 of the periodic table. These elements are called alkaline earth metals.

The reactions of group 2 metals with water.
Magnesium only reacts well with water when heated in steam.

$$Mg(s) + H_2O(g) \rightarrow MgO(s) + H_2(g)$$

Calcium reacts vigorously with cold water:

$$Ca(s) + 2H_2O(l) \rightarrow Ca(OH)_2(aq) + H_2(g)$$

Barium reacts more vigorously than calcium:

$$Ba(s) + 2H_2O(l) \rightarrow Ba(OH)_2(aq) + H_2(g)$$

(a) **(i)** How does the reactivity of alkaline earth metals change down group 2? **[1]**

(ii) Explain this difference in reactivity. Use ideas of atomic structure in your answer. **[4]**

(b) **(i)** Write the formula of calcium chloride. **[1]**

(ii) Describe how a sample of calcium chloride could be produced from calcium hydroxide. **[5]**

2. Some lumps of zinc (5 g) were put into a flask and 100 cm^3 of hydrochloric acid (100 g/dm^3) added. The temperature was 20 °C.

How would the rate of formation of hydrogen be affected in each of the following changes made in turn, all other conditions remaining the same? Explain your answers.

In each experiment the acid is in excess. **[8]**

New condition	Change, if any	Explanation
Use 5 g of powdered zinc		
Use 40 °C		
Use 100 cm^3 of hydrochloric acid (50 g/dm^3)		
Use 100 cm^3 of ethanoic acid (100 g/dm^3)		

Extension material

Check in the table on pages 4 and 5 to see which sections you need to study.

Chapter	Section	Studied in class	Revised	Practice questions
4 *Water*	**Salt from sea water**			
	Testing for water			
	Hardness of water			
	Soaps and soapless detergents			
	Solubility of gases in water			
	Solubility of solids			
	Colloids and emulsions			
	Purification of water supplies			
5 *Acids, bases and salts*	**Acids**			
	Neutralisation			
	Preparing soluble salts			
	Preparing insoluble salts			
	Testing for ions			
6 *Metals and redox*	**Reactivity of aluminium**			
	Anodising aluminium			
	Redox reactions			
	Rusting and its prevention			
	Cells			
	Alloys			
7 *Further carbon chemistry*	**Hardening natural oils**			
	Isomerism			
	Alcohols			
	Ethanoic acid			
	Esters			
	Further polymers			
8 *Further quantitative chemistry*	**The mole**			
	Calculating formulae from percentages			
	Volume changes in chemical reactions			
	Concentration of solutions			
9 *Electrochemistry and electrolysis*	**Electrical conductivity**			
	Electrolysis of molten zinc chloride			
	Electrolysis of aqueous solutions			
	Uses of electrolysis			
	Quantitative electrolysis			
10 *Collection of gases*	**Methods of collecting gases**			
	Tests for gases			
11 *Food and drugs*	**Carbohydrates, proteins and fats**			
	Vitamins and food additives			
	Drugs			
12 *Radicals*	**How radicals form**			
	Depletion of the ozone layer			
13 *Group 2 metals*	**Physical properties of alkaline earth metals**			
	Reactions with air and water			
	Explaining the difference in reactivity			

Chapter

4 Water

The following topics are covered in this section:

- *Salt from sea water*
- *Testing for water*
- *Hardness of water*
- *Soaps and soapless detergents*
- *Solubility of gases in water*
- *Solubility of solids*
- *Colloids and emulsions*
- *Purification of water supplies*

LEARNING SUMMARY

After studying this section you should be able to:

- *describe the tests for water*
- *recall that hard water does not lather well with soap but forms scum*
- *explain what causes temporary hard water and permanent hard water, what are the advantages and disadvantages of hard water and how hardness can be removed*
- *describe how soaps and soapless detergents act as cleaning agents*
- *describe how the solubility of gases and solids changes with increasing temperature*
- *recall how a safe water supply can be produced.*

Water is a very important substance as it is essential for so many fundamental processes. It is, however, very difficult to get pure because is so good at dissolving other substances. This ability to dissolve other things has important consequences.
A clean water supply is essential for good health. Many diseases e.g. cholera are a consequence of impure water supplies

Salt from sea water

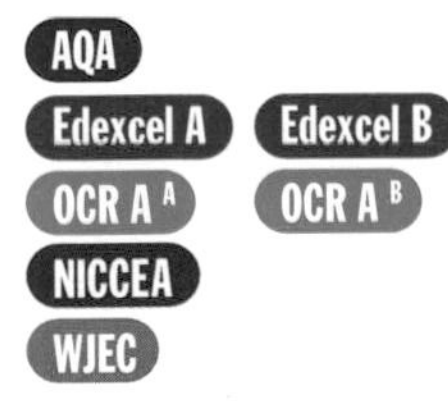

Evaporation occurs at a surface and the larger the surface the faster the evaporation. Evaporation needs energy to take place.

Sea water contains a large amount of dissolved material (2.5). The most common ions in sea water are sodium ions and chloride ions.

When a sample of sea water is evaporated to half its original volume and left to cool, solid sodium chloride crystallises out.

The sea has been a traditional source of salt, which has been used for flavouring food but more importantly, before refrigeration, for preserving it.

In warm countries large beds of sea water are left to evaporate using energy from the sun. These beds cover a large area and are shallow.

Testing for water

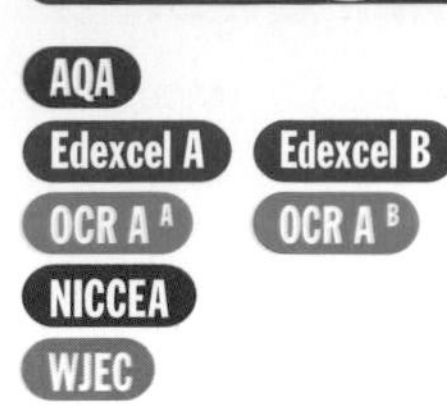

The presence of water in a liquid can be shown using anhydrous copper(II) sulphate or cobalt(II) chloride paper.

The results are shown in **Table 4.1** below:

Chemical used to detect water	Colour before testing	Colour after testing
Anhydrous copper(II) sulphate	white	blue
Cobalt(II) chloride paper	blue	pink

Candidates frequently get the colours wrong here especially as blue is the correct answer before with one chemical but after with the other.

A positive test with one of these chemicals shows that water is present: it does not show pure water. A solution of sodium chloride would give a positive test. To show that a liquid is pure water, melting and boiling point tests should be carried out.

Pure water freezes at 0°C and boils at 100°C.

Hardness of water

What is hard water?

Distilled water is pure water and contains no dissolved solid impurities.

Rain water closely resembles distilled water but contains some dissolved carbon dioxide.

Both distilled water and rain water lather well with soap solution without producing any scum.

Rain water contains a weak acid – carbonic acid H_2CO_3.

Some water samples, however, do not lather well and form scum. These samples are called hard water.

Distilled water and rain water are **soft** water.

Hard water is caused by certain dissolved substances in water. These substances become dissolved in the water as it trickles through rocks.

Calcium sulphate is very slightly soluble in water. When water trickles through deposits of gypsum (calcium sulphate) a small amount of the calcium sulphate dissolves.

Calcium carbonate is virtually insoluble in water but rain water, containing dissolved carbon dioxide, does react with some of the calcium carbonate forming **calcium hydrogencarbonate.**

$$CaCO_3(s) + H_2O(l) + CO_2(g) \rightleftharpoons Ca(HCO_3)_2\,(aq)$$

calcium carbonate + water + carbon dioxide ⇌ calcium hydrogencarbonate

Hard water is caused by dissolved calcium and magnesium compounds in water.

There are two types of hardness in water – **temporary hardness** and **permanent hardness**.

Temporary hardness is caused by dissolved **calcium hydrogencarbonate**. This can be removed by boiling. Heating the water decomposes the calcium hydrogencarbonate.

$$Ca(HCO_3)_2\,(aq) \rightleftharpoons CaCO_3(s) + H_2O(l) + CO_2(g)$$

Chemical descalers contain a weak acid that reacts with calcium carbonate.

Much of this calcium carbonate forms the scale inside kettles in hard water areas.

Permanent hardness is caused by dissolved **calcium sulphate** and **magnesium sulphate**.

This hardness is not removed by boiling but has to be removed by chemical treatment.

Table 4.2 below gives the advantages and disadvantages of hard water.

Disadvantages of hard water	Advantages of hard water
Forms scale or deposits in kettles and on heating elements making them less efficient and increasing costs. Clogs up pipes.	Hard water is better for brewing beer. It is for this reason that Burton-on-Trent became a centre for beer brewing.
Wastes soap as more soap is needed to lather.	Supplies calcium which is essential for development of bones especially in children.
Forms scum which can damage the finish on garments.	Helps to reduce heart illnesses.

Removal of hardness in water

All hardness in water can be removed from water by distillation. This is not, however, an economic process on a large scale.

Temporary hardness is removed by boiling.

This decomposes calcium hydrogencarbonate and removes the hardness. Boiling does not remove permanent hardness.

Washing soda softens all types of hard water.

KEY POINT **Permanent and temporary hardness can be removed with sodium carbonate crystals (washing soda).**

$$Ca(HCO_3)_2(aq) + Na_2CO_3(aq) \rightarrow CaCO_3(s) + 2NaHCO_3(aq)$$

calcium hydrogencarbonate + sodium carbonate → calcium carbonate + sodium hydrogencarbonate

$$CaSO_4(aq) + Na_2CO_3(aq) \rightarrow CaCO_3(s) + Na_2SO_4(aq)$$

calcium sulphate + sodium carbonate → calcium carbonate + sodium sulphate

$$MgSO_4(aq) + Na_2CO_3(aq) \rightarrow MgCO_3(s) + Na_2SO_4(aq)$$

magnesium sulphate + sodium carbonate → magnesium carbonate + sodium sulphate

Using a water softener can save money. Scale on heating elements make them less efficient. Hard water also uses more soap.

In each case calcium or magnesium carbonate is precipitated and this removes the hardness from the water.

Other substances, e.g. calcium hydroxide, 'Calgon' (sodium metaphosphate) and sodium sesquicarbonate work in a similar way by precipitating the substances that cause hardness.

Commercial water filters contain an ion exchanger resin to soften water and activated charcoal to remove chlorine and metal ions.

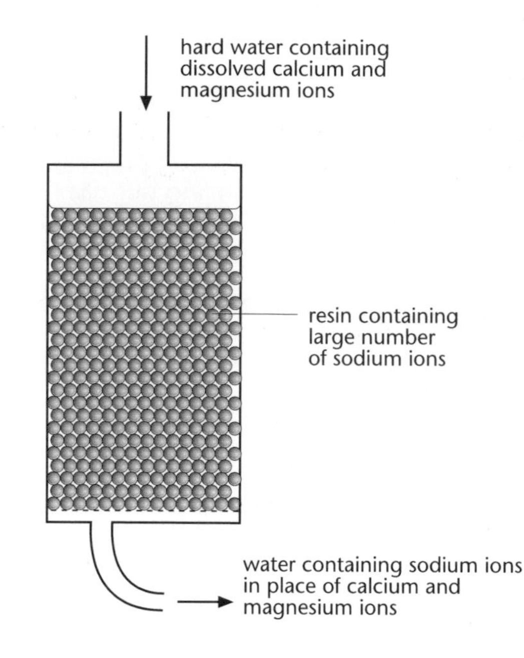

Hardness can be removed by using an **ion-exchange column (Fig. 4.1)**. A column is filled with a suitable resin in small granules. The resin has an excess of sodium ions. As the hard water trickles through the column, calcium and magnesium ions in the water are exchanged for sodium ions. The resulting water contains no calcium and magnesium ions and therefore has no hardness. When all the sodium ions are used up, the column can be recharged.

Fig. 4.1

PROGRESS CHECK

1. Which metal ions in solution cause hardness in water?
2. Write down the name of the substance that causes temporary hardness.
3. Write down the names of two substances that cause permanent hardness.
4. Temporary hardness can be removed by adding the correct amount of calcium hydroxide. Write a symbol equation for the reaction taking place.
5. Why is it important when softening hard water with calcium hydroxide not to use too much?

1. Calcium and magnesium; 2. Calcium hydrogencarbonate; 3. Calcium sulphate and magnesium sulphate; 4. $Ca(HCO_3)_2 + Ca(OH)_2 \rightarrow 2CaCO_3 + 2H_2O$; 5. Adding too much calcium hydroxide puts more calcium ions into the water and therefore more hardness.

Soaps and soapless detergents

OCR A B

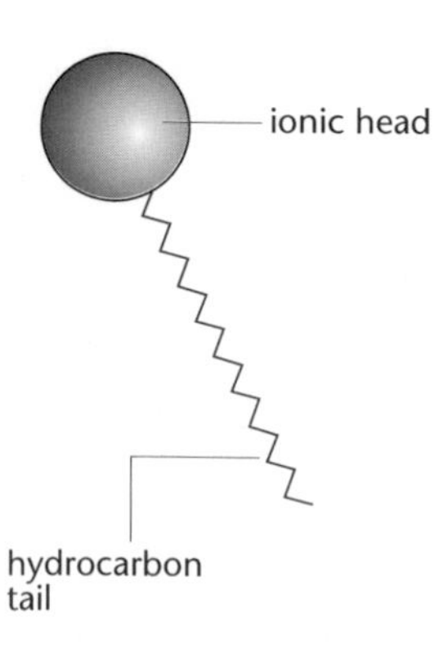

Fig. 4.2 Detergent molecule

The word detergent means 'cleaning agent'. Soap is one type of detergent.

Other soapless detergents have been developed.

Soaps and soapless detergent molecules have a similar structure (Fig. 4.2). In each case there are two parts to the molecule – an 'ionic' head and a long hydrocarbon tail.

The ionic head is said to be **hydrophilic** or water loving and the hydrocarbon tail – **hydrophobic** or water hating.

Soap

Soap is produced by heating vegetable or animal fats and oils with concentrated sodium hydroxide solution or potassium hydroxide solution (an alkali).

Natural fat or oil + sodium hydroxide → sodium octadecanoate + glycerol.

Sodium octadecanoate (sometimes called sodium stearate) acts as a soap.

It has a formula $C_{17}H_{35}COO^- Na^+$. The $C_{17}H_{35}$– represents the hydrophobic chain and the $-COO^-$ the hydrophilic ionic group.

The process of making soap is sometimes called **saponification**.

Soapless detergents

In the past 50 years there has been a move from soaps to soapless detergents.

Soapless detergents are produced using some of the long chain residues from fractional distillation of crude oil.

Soapless detergents are made by reacting crude oil residues with concentrated sulphuric acid.

Crude oil residue + conc. sulphuric acid → soapless detergent + water

Soapless detergent molecules have a hydrocarbon tail and an ionic head, which is an $-SO_3^-$ head.

Cleaning action of soaps and soapless detergents

Fig. 4.3 below summarises the cleaning action of a detergent (soap or soapless).

Soaps and soapless detergents can be solid or liquid.

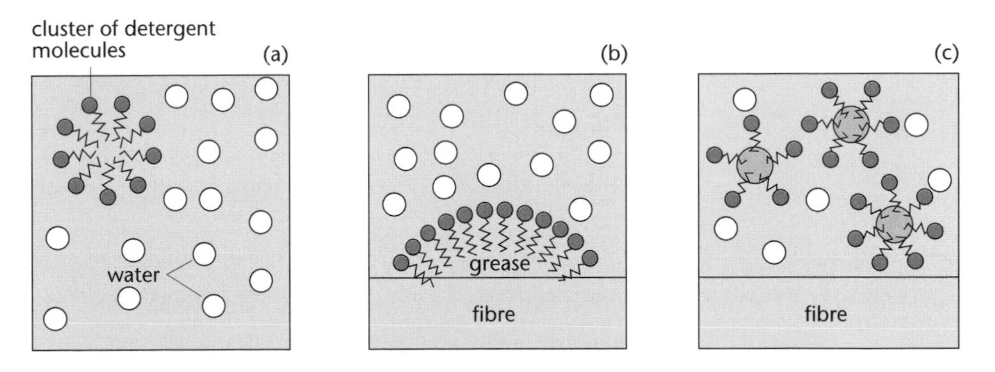

Picture (a) shows a detergent in water. The tails of the molecules bunch together and try to keep away from water molecules.

Picture (b) The tails of the detergent molecules stick into the grease on the surface of the cloth. The water molecules pull on the heads of the detergent molecules. The grease is lifted from the cloth. Stirring the mixture aids this process.

Picture (c) The grease is suspended in the solution and the repulsive forces between the heads on the detergent molecules stop the grease going back on the cloth.

Advantages of soapless detergents

Soaps lather well in soft water but less well in hard water. Soapless detergents lather well in hard and soft water.

In hard water soap reacts with dissolved calcium and magnesium ions in the water and forms **scum**.

Calcium ions are precipitated and so water is no longer hard.

$$2C_{17}H_{35}COO^- Na^+(aq) + Ca^{2+}(aq) \rightarrow (C_{17}H_{35}COO)_2Ca(s) + 2Na^+(aq)$$

sodium octadecanoate(soap) + calcium ions →
calcium octadecanoate (scum) + sodium ions

PROGRESS CHECK

1. Which metal ions cause hardness in water?
2. Which substance in water causes temporary hardness?
3. Why does this compound get into water in limestone areas?
4. What is seen when soap is added to hard water?
5. What is seen when a soapless detergent is added to hard water?
6. What is used to make soap from a natural oil?
7. What is used to make a soapless detergent from crude oil residues?

1. Calcium and magnesium; 2. Calcium hydrogencarbonate; 3. Rain water containing dissolved carbon dioxide (carbonic acid) reacts with limestone to form calcium hydrogen carbonate; 4. Scum and little lather; 5. No scum and a lot of lather; 6. Sodium or potassium hydroxide solution (strong alkali); 7. Concentrated sulphuric acid.

Solubility of gases in water

AQA
OCR A [A] OCR A [B]
NICCEA

Many gases dissolve in water but they dissolve to very different extents. Only two gases, hydrogen chloride and ammonia, are very soluble in water.

In both cases the gases actually react with the water to form new substances. Hydrogen chloride dissolves in water to form the strong acid – **hydrochloric acid**.

Ammonia dissolves in water to form the alkali, **ammonium hydroxide.**

$$NH_3 + H_2O \rightarrow NH_4OH$$

Carbon dioxide and sulphur dioxide dissolve less to form the weak acids, **carbonic acid** and **sulphuric (IV) acid.**

Sulphuric(IV) acid is sometimes called sulphurous acid.

$$CO_2 + H_2O \rightleftharpoons H_2CO_3$$

$$SO_2 + H_2O \rightleftharpoons H_2SO_3$$

Chlorine dissolves in water forming chlorine water – a mixture of hydrochloric acid and chloric(I) acid. This solution is used as a bleaching agent.

Chloric(I) acid is sometimes called hypochlorous acid.

$$Cl_2 + H_2O \rightarrow HCl + HOCl$$

Oxygen is only very slightly soluble in water. However, this dissolved oxygen is essential for fish and other living material in rivers. Any pollution of the water reduces the amount of dissolved oxygen in water.

The volume of gas dissolved in water increases with increasing pressure and decreases with increasing temperature.

1. How could you distinguish between a solution of ammonia in water and a solution of hydrogen chloride in water?
2. Put these three gases in order of decreasing solubility in 100 cm^3 of water under the same conditions of temperature and pressure.
   oxygen ammonia carbon dioxide
3. Which of the following would dissolve the most gas?
   (a). High temperature and high pressure.
   (b). High temperature and low pressure.
   (c). Low temperature and high pressure.
   (d). Low temperature and low pressure.
4. Waste water from a power station enters a river. The water is hot. Why can this cause distress to fish in the river? (Relate your answer to the solubility of gases.)
5. When the top is removed from a bottle of fizzy lemonade, bubbles of gas escape from the liquid. Why does this happen?

1. Add litmus (or other indicator). Solution of ammonia is blue. Solution of hydrogen chloride is red; 2. Ammonia carbon dioxide oxygen; 3. (c); 4. Less oxygen in dissolved in the river when the water gets hotter; 5. When pressure is released less carbon dioxide will dissolve and excess gas escapes.

Solubility of solids

AQA
OCR A[A] OCR A[B]
NICCEA

Water dissolves a wide range of different substances. Water is said to be a good solvent and the substances being dissolved are called solutes.

Because water is such a good solvent, it is difficult to keep it pure.

Water is a **polar solvent**, i.e. it contains small positive and negative charges caused by the slight movement of electrons in covalent bonds. Polar solvents dissolve compounds containing ionic bonds, e.g. sodium chloride. **Non-polar solvents** (e.g. tetrachloromethane, CCl_4) are poor at dissolving ionic compounds but are good solvents for molecular compounds.

A solution that contains as much solute as can be dissolved at a particular temperature is called a saturated solution.

If any more solute is added to a saturated solution, the extra solute remains undissolved.

When quoting a solubility you must also quote the temperature.

The solubility of a solute is the mass of the solute (in grams) that dissolves in 100g of solvent at a particular temperature to form a saturated solution.

Generally, the solubility of ionic compounds in water (e.g. potassium nitrate) increases as temperature rises.

Fig. 4.4 below shows graphs of the solubilities of three ionic compounds in 100g of water at different temperatures. These graphs are called **solubility curves.**

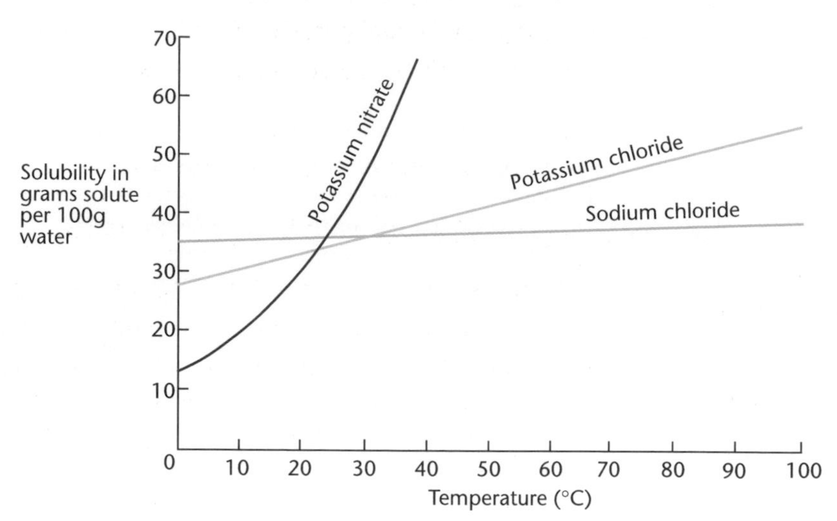

Table 4.3 gives the solubility of copper(II) sulphate in 100g of water at different temperatures.

Temperature in °C	0	10	20	40	60	80
Solubility in g/100g water	14	17	21	29	40	55

The solubility of copper(II) sulphate clearly increases with increasing temperature. If a saturated solution of copper(II) sulphate containing 100g of water was cooled from 80°C to 20°C, some of the copper(II) sulphate that dissolved at 80°C will no longer dissolve at 20°C. This extra copper(II) sulphate **crystallises**.

At 80°C 55g dissolves in 100g of water and at 20°C 21g of copper(II) sulphate dissolves. This means 34g of copper(II) sulphate crystals will be formed.

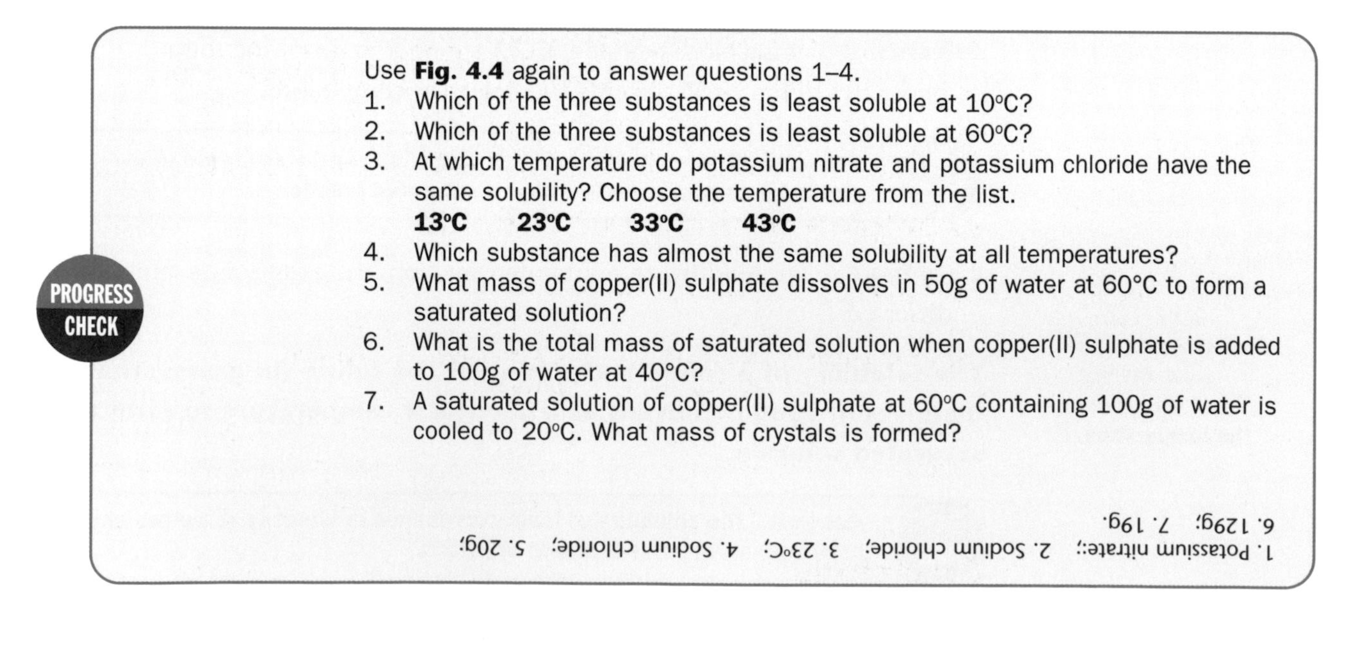

PROGRESS CHECK

Use **Fig. 4.4** again to answer questions 1–4.

1. Which of the three substances is least soluble at 10°C?
2. Which of the three substances is least soluble at 60°C?
3. At which temperature do potassium nitrate and potassium chloride have the same solubility? Choose the temperature from the list.
   13°C **23°C** **33°C** **43°C**
4. Which substance has almost the same solubility at all temperatures?
5. What mass of copper(II) sulphate dissolves in 50g of water at 60°C to form a saturated solution?
6. What is the total mass of saturated solution when copper(II) sulphate is added to 100g of water at 40°C?
7. A saturated solution of copper(II) sulphate at 60°C containing 100g of water is cooled to 20°C. What mass of crystals is formed?

1. Potassium nitrate; 2. Sodium chloride; 3. 23°C; 4. Sodium chloride; 5. 20g; 6. 129g; 7. 19g.

Colloids and emulsions

OCR A B

In a sugar **solution** the particles of the solute are distributed through the solvent as individual particles. Individual sugar particles cannot be seen. Light passes through the solution i.e. it is transparent.

In a mixture of sand and water, called a **suspension**, individual particles of sand can be seen. On standing, the particles of sand settle to the bottom.

There is a stage between these two extremes:

Colloids are also mixtures of two substances, with one substance distributed through the other but the particles distributed or **dispersed** are much larger and can be seen with a microscope. Light does not pass through the mixture.

A colloid cannot be separated by filtration. Co-agulation has to take place.

Fig. 4.5 shows a suspension. The mixture has a **disperse phase** and a **continuous phase.**

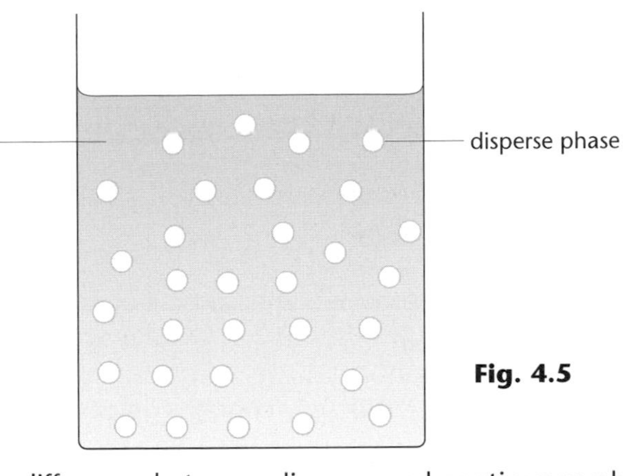

Fig. 4.5

Table 4.4 shows the difference between disperse and continuous phases.

Phase	What is it?	It can be
Disperse phase	These are the droplets or bubbles distributed throughout the mixture.	Tiny particles of solid, tiny droplets of liquid or tiny bubbles of gas.
Continuous phase	The material through which the droplets or bubbles are spread.	A liquid, a solid or a gas.

Examples of different colloids

You can find many examples of colloids in the home.

Table 4.5 gives examples of different colloids.

Name of colloid	Disperse phase	Continuous phase	Example
Emulsion	liquid	liquid	Milk, butter, salad cream
Foam	gas	liquid	Hair mousse, whipped cream
Sol	solid	liquid	Paint

Emulsions

Oil and water form two separate layers with the oil floating on the water. If the mixture is shaken vigorously one mixture is formed with droplets of oil distributed throughout the watery layer. On standing, the mixture separates

A soap or soapless detergent acts as an emulsifying agent when cleaning action of detergents is considered. (Fig.4.6)

again into two layers. If a substance called an **emulsifying agent** or **emulsifier** was present the mixture would not separate and an emulsion would be formed. **Fig. 4.6** shows how the emulsifying agent prevents the droplets of oil coming together again. **Fig. 4.7** shows some milk under the microscope. It consists of small bubbles of fat (disperse phase) spread through a watery or aqueous layer (continuous phase).

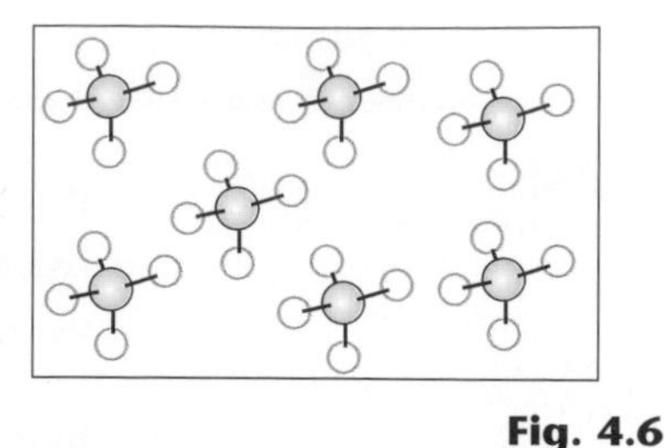

Fig. 4.6

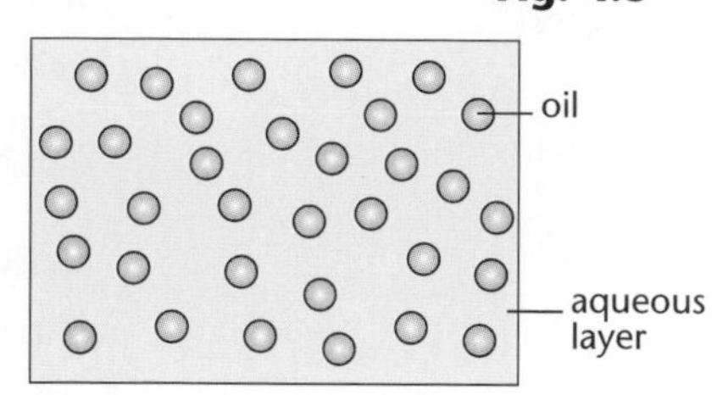

Fig. 4.7

This is called an **oil-in-water emulsion**.

Be clear in the difference between an oil-in-water emulsion and a water-in-oil emulsion.

If a water-soluble dye is added the continuous phase is coloured but not the disperse phase. The result is shown in **Fig. 4.8.**

When the cream of the milk is agitated for a long time it is turned into butter. Butter is a **water-in-oil emulsion**. Droplets of water are distributed through the oily layer.

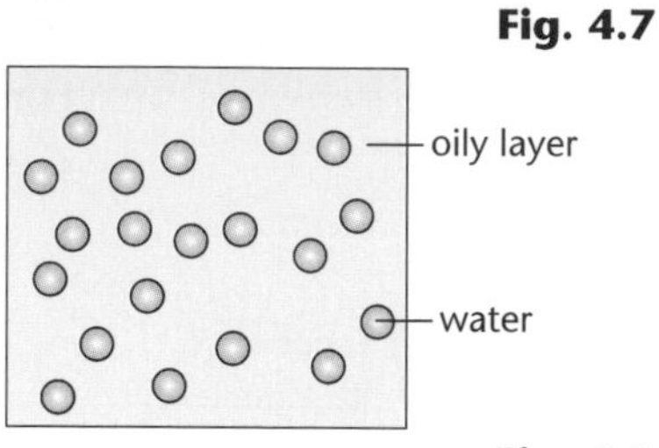

Fig. 4.8

Gel

A gel is a special type of colloid. It contains large molecules which form a web structure which traps water. The result is a very thick liquid or solid mixture.

A fruit jelly contains gelatin with added fruit flavouring. When it is added to hot water the gelatin dissolves. On cooling, the gelatin sets into a large structure with water trapped within it.

Another example is wallpaper paste. This is largely starch. On mixing with water a liquid is formed which quickly thickens to form a semi-solid paste.

Coagulation of a colloid

The droplets in the disperse phase of a colloid are often charged with the same charge. Repulsion between the droplets prevents them **coagulating**. **Fig. 4.9** shows a colloid where the droplets are negative.

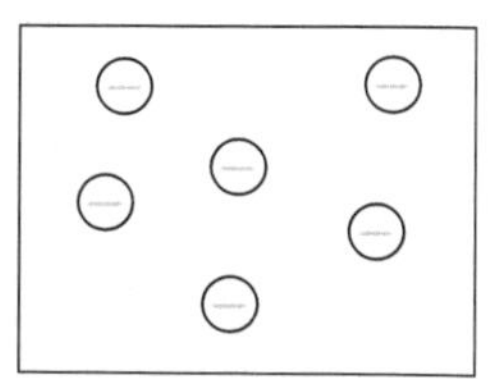

Fig. 4.9

With a positively charged colloid, sulphate ions will be better than chloride ions for coagulating the colloid.

If this charge is removed coagulation can take place and a precipitate will form. Using solutions of equal molar concentrations of sodium sulphate (containing Na^+ ions), magnesium sulphate (containing Mg^{2+} ions) and aluminium sulphate (containing Al^{3+} ions), it can be shown that aluminium ions are more effective than magnesium ions and magnesium ions are more effective than sodium ions.

KEY POINT **More highly charged ions are more effective at coagulating colloids.**

PROGRESS CHECK

1. What would be seen if a water soluble dye was added to butter under a microscope?
   Choose words from this list:
   emulsion **foam** **gel**
2. Which of the words in the list describes a hand cream?
3. Which of the words in the list describes a honeycomb bar?
4. How could you show the difference between a true solution and an emulsion?
5. Describe the structure of an emulsifying agent molecule.

1. The disperse phase is coloured but not the continuous phase. 2. Emulsion; 3. Foam; 4. Light passes through a true solution but is dispersed by an emulsion. 5. An emulsifying agent molecule consists of an ionic head and a hydrocarbon chain (e.g. detergent molecule).

Purification of water supplies

AQA
OCR A [B]
WJEC

A clean, safe supply of water provided to homes and industry is essential for public health.

As a country becomes more developed, more water is required.

Every day each person in the United Kingdom uses about 120 litres of water – for washing, flushing toilets, cooking, etc. Industry too uses a lot of water – it takes 26 000 litres of water to make one tonne of newsprint, 45 500 litres to make a tonne of steel and up to 7 litres to make a pint of beer.

Producing tap water

Carbon in the form of charcoal is a good absorber of unwanted colours and tastes.

There are various essential steps in producing tap water:

- water is taken from a clean river or reservoir
- the water is passed through a screen (sieve) to remove solid objects
- it is left to stand to allow solid material to settle out
- it is then filtered through a gravel bed to remove impurities in suspension
- it is then **chlorinated**. Chlorine is added in small amounts to kill bacteria and other harmful micro-organisms.

If the water contains iron compounds it can make tea have an inky colour and a bitter taste. It can also cause brown stains on clothes after washing.

Other forms of treatment may be used in special cases.

These include:

- adding aluminium sulphate to coagulate colloidal clay in water
- using a mixture of carbon and water to remove tastes and odours from the water
- using lime to correct acidity of the water
- adding sulphur dioxide to remove excess chlorine.

Aluminium sulphate contains Al^{3+} which are very efficient at coagulation.

The water produced at the end of all these processes is not pure water. It is water that is safe to drink.

Fluoride is added to water to improve the dental health of consumers. Some people are against adding substances to water. Large concentrations of sodium fluoride are used as a rat poison.

The water leaving a sewage works is pure enough to be used again. Water in the River Thames is used several times for water supply before it enters the sea.

Recycling waste water

After water has been used by a home or a factory, it must be cleaned up before being returned to rivers and possibly re-used for water supply.

This is done in a sewage works. The treatment of waste water involves:

- filtering it to remove solid materials
- using bacteria to break down the waste.

PROGRESS CHECK

1. Why is chlorine bubbled through water during treatment?
2. Which process is used to remove large objects such as leaves from water?
3. Which process removes suspended solids from water?
4. In the recycling of water, waste water is sprayed onto a large bed of gravel where much of the waste is broken down. What breaks down the waste?

1. To kill harmful bacteria or other micro-organisms; 2. Sieving or screening;
3. Settling; 4. Bacteria.

Sample GCSE question

1. An experiment was carried out to compare the hardness of four water samples labelled A, B, C and D.

25 cm^3 of water sample A was transferred to a conical flask. Soap solution was then added to the flask in small portions using a burette until a lasting lather was formed. The experiment was repeated with water samples B, C and D. Fresh specimens of each water sample were then boiled and tested again with soap solution.

The results are shown in the table.

Water sample	Volume of soap solution required before boiling (cm^3)	Volume of soap solution required after boiling (cm^3)
A	5.0	5.0
B	1.0	1.0
C	11.0	6.0
D	9.0	1.0

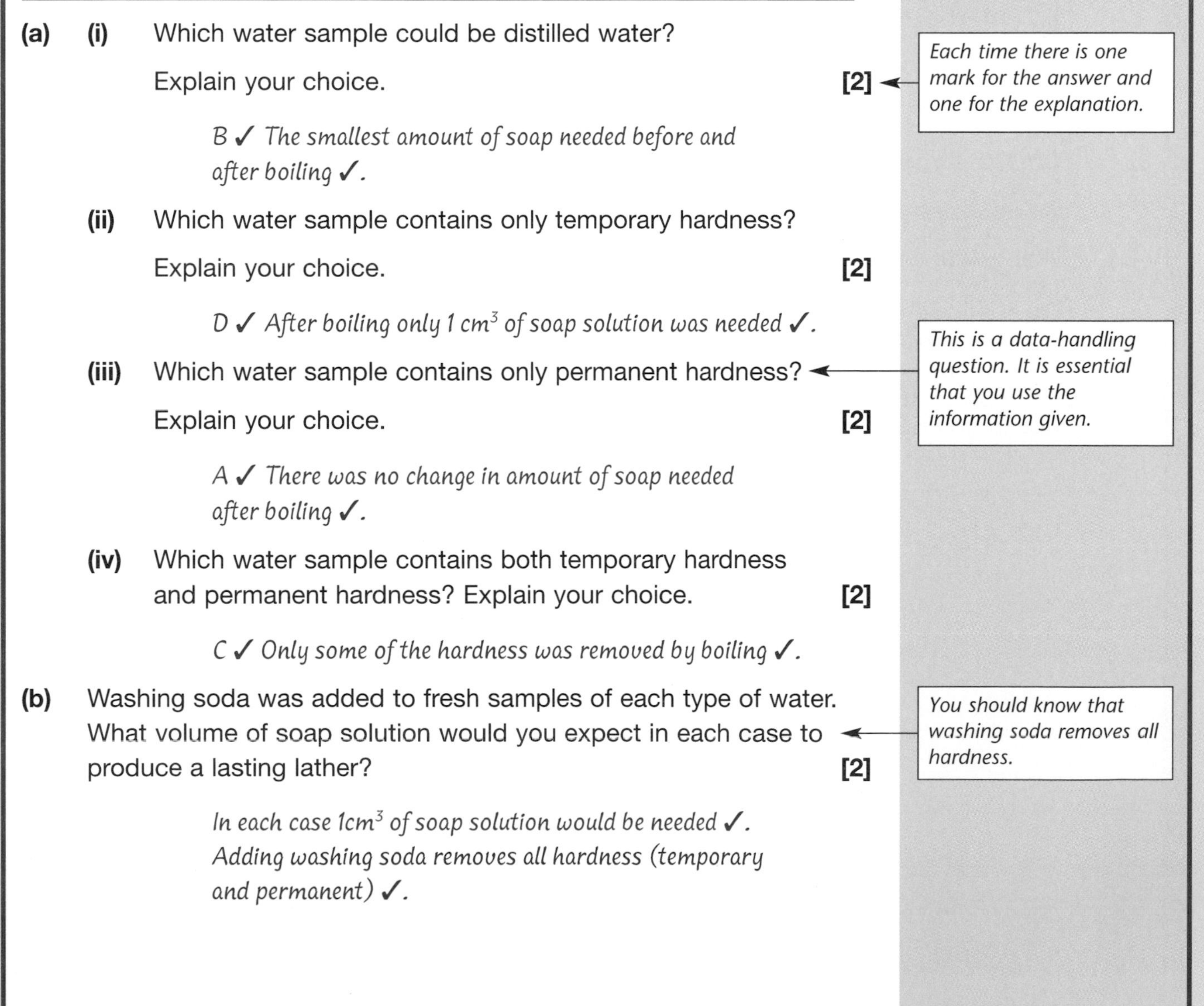

(a) **(i)** Which water sample could be distilled water?

Explain your choice. **[2]**

Each time there is one mark for the answer and one for the explanation.

B ✓ The smallest amount of soap needed before and after boiling ✓.

(ii) Which water sample contains only temporary hardness?

Explain your choice. **[2]**

D ✓ After boiling only 1 cm^3 of soap solution was needed ✓.

(iii) Which water sample contains only permanent hardness?

This is a data-handling question. It is essential that you use the information given.

Explain your choice. **[2]**

A ✓ There was no change in amount of soap needed after boiling ✓.

(iv) Which water sample contains both temporary hardness and permanent hardness? Explain your choice. **[2]**

C ✓ Only some of the hardness was removed by boiling ✓.

(b) Washing soda was added to fresh samples of each type of water. What volume of soap solution would you expect in each case to produce a lasting lather? **[2]**

You should know that washing soda removes all hardness.

In each case 1cm^3 of soap solution would be needed ✓. Adding washing soda removes all hardness (temporary and permanent) ✓.

Exam practice questions

1. The table gives the solubilities of potassium nitrate, in g per 100g of water, at different temperatures.

Temperature in °C	0	10	20	40	60	80	100
Potassium nitrate	13	21	32	64	110	169	246

(a) Plot this data on a grid and draw a solubility curve for potassium nitrate. **[3]**

(b) From the graph find:

(i) the solubility of potassium nitrate at 30°C **[1]**

(ii) the maximum number of grams of potassium nitrate that would dissolve in 50g of water at 50°C **[1]**

(c) A solution is made from 40g of potassium nitrate dissolved in 100g of water at 60°C. It is then cooled.

(i) At what temperature would this be a saturated solution? **[1]**

(ii) What would happen if this solution was cooled to 10°C? **[2]**

2. Sam finds a colourless liquid. She wants to find out if this liquid contains water.

She adds anhydrous copper(II) sulphate to the liquid.

(a) What colour change would she expect if the liquid contains water? **[2]**

She adds Universal Indicator to the liquid and it turns green (pH 7). She evaporates a sample to dryness when a white residue is left. The liquid boils at 102°C.

(b) Which of her observations suggest that the liquid could be pure water and which suggest the liquid could be an aqueous solution? **[2]**

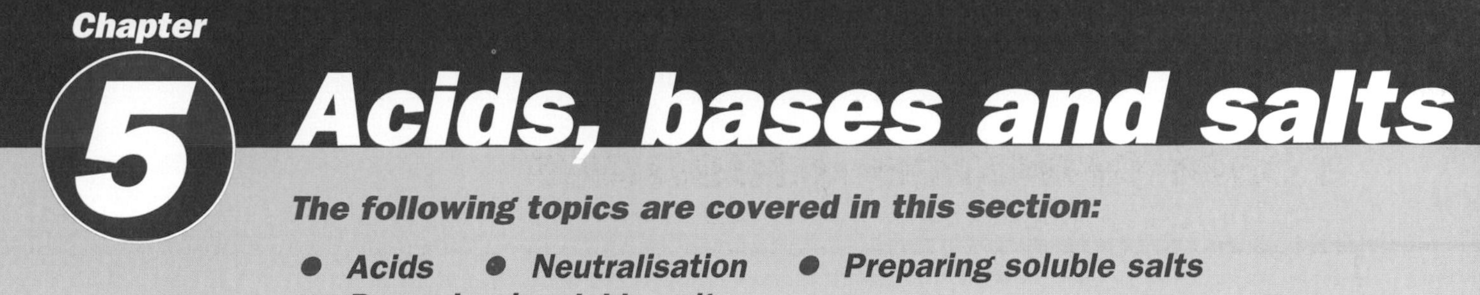

Chapter 5 Acids, bases and salts

The following topics are covered in this section:

- ***Acids***
- ***Neutralisation***
- ***Preparing soluble salts***
- ***Preparing insoluble salts***

LEARNING SUMMARY

After studying this section you should be able to:

- ***recall the properties of acids***
- ***explain the difference between strong and weak acids***
- ***explain the process of neutralisation and give common examples***
- ***describe how soluble and insoluble salts can be prepared***
- ***recall tests for common positive and negative ions.***

KEY POINT

You will know that there are substances called indicators that can identify acids and alkalis by changing colour. Litmus, for example, turns red in acids and blue in alkalis.
Acids and alkalis react in neutralisation reactions. This Unit studies the topic at greater depth. It includes tests for common positive and negative ions.

Acids

KEY POINT

Acids:

- **are compounds containing hydrogen that can be replaced by a metal**
- **dissolve in water to form hydrogen, H^+ ions**
- **are proton donors.**

An acid with one replaceable hydrogen atom is called a monobasic acid.

The hydrogen in hydrochloric acid, HCl, can be replaced by sodium to form sodium chloride, NaCl.

Table 5.1 below gives a list of names and formulae of some common acids.

acid	formula	
Hydrochloric acid	HCl	
Sulphuric acid	H_2SO_4	Mineral acids
Nitric acid	HNO_3	
Ethanoic acid (Acetic acid)	CH_3COOH	Contained in vinegar
Ethanedioic acid	$C_2O_4H_2$	Contained in rhubarb leaves
Citric acid	$C_3H_8O_7$	Contained in lemon juice

Ethanoic acid contains four hydrogen atoms per molecule but is a monobasic acid.

Only one hydrogen atom in ethanoic acid can be replaced, forming sodium ethanoate CH_3COONa.

Properties of acids

Although there are a large number of different acids, there are a number of general chemical reactions common to all acids.

Indicators

Acids turn indicators to their characteristic colours, e.g. litmus turns red.

Fairly reactive metals

Acids react with fairly reactive metals (e.g. magnesium and zinc) to form a salt and hydrogen gas.

The equations show state symbols. These are optional unless they are specifically asked for.

e.g. $Mg(s) + H_2SO_4(aq) \rightarrow MgSO_4(aq) + H_2(g)$

magnesium + sulphuric acid → magnesium sulphate + hydrogen

$Zn(s) + 2HCl(aq) \rightarrow ZnCl_2(aq) + H_2(g)$

Zinc + hydrochloric acid → zinc chloride + hydrogen

Nitric acid is an exception. This acid tends to release oxides of nitrogen when it reacts with metals.

Metal oxides

Acids react with metal oxides to form a salt and water only.

Metal oxides are called bases.

In most cases warming is necessary.

e.g. $CuO(s) + H_2SO_4(aq) \rightarrow CuSO_4(aq) + H_2O(l)$

copper(II) oxide + sulphuric acid → copper(II) sulphate + water

$ZnO(s) + 2HCl(aq) \rightarrow ZnCl_2(aq) + H_2O(l)$

Zinc oxide + hydrochloric acid → zinc chloride + water

A blue solution is formed with Copper Sulphate.

Metal carbonates

Acids react with carbonates (or hydrogencarbonates) to form carbon dioxide, a salt and water.

e.g. $CaCO_3(s) + 2HCl(aq) \rightarrow CaCl_2(aq) + H_2O(l) + CO_2(g)$

calcium carbonate + hydrochloric acid → calcium chloride + water + carbon dioxide

$2NaHCO_3(s) + H_2SO_4(aq) \rightarrow Na_2SO_4(aq) + 2H_2O(l) + 2CO_2(g)$

sodium hydrogencarbonate + sulphuric acid → sodium sulphate + water + carbon dioxide

Metal hydroxides

Alkalis react with acids to form a salt and water only.

Metal hydroxides that dissolve in water are called alkalis.

These tests are used to identify the presence of an acid.

e.g. $NaOH(aq) + HCl(aq) \rightarrow NaCl(aq) + H_2O(l)$

sodium hydroxide + hydrochloric acid → sodium chloride + water

These reactions of acids can be represented by **ionic equations**.

You may be expected to complete ionic equations to score marks targeted at A or A*.

e.g. $Mg(s) + 2H^+(aq) \rightarrow Mg^{2+}(aq) + H_2(g)$

$O^{2-}(s) + 2H^+(aq) \rightarrow H_2O(l)$

$CO_3^{2-}(s) + 2H^+(aq) \rightarrow H_2O(l) + CO_2(g)$

$H^+(aq) + OH^-(aq) \rightarrow H_2O(l)$

Strong and weak acids

Dry acids do not show acidic properties. Water must be present.

Some acids completely **ionise** when they dissolve in water. These are called **strong acids**.

The idea of dissociation of acids came from Arrhenius. These ideas were developed later by Lowry and Bronsted.

A solution of a strong acid will have a high concentration of hydrogen ions, H^+.

e.g. sulphuric acid

$H_2SO_4(l) \xrightarrow{water} 2H^+(aq) + SO_4^{2-}(aq)$

Arrhenius' theory did not involve the solvent and so could not explain some examples.

Other acids do not completely ionise on dissolving in water. Some of the molecules remain **un-ionised** in the solution. These are called **weak acids**.

e.g. $CH_3COOH(l) \overset{water}{\rightleftharpoons} CH_3COO^-(aq) + H^+(aq)$

In a solution of ethanoic acid (1 mol/dm^3) there are about four molecules ionised in every thousand.

A strong acid is one that is completely ionised. It is nothing to do with how corrosive it is. Nor is it a measure of the concentration of an acid.

PROGRESS CHECK

1. Which of the acids in the list is not a strong acid?
   ethanoic acid **hydrochloric acid** **nitric acid** **sulphuric acid**
2. Which acid reacts to form sulphates?
3. Which acid reacts to form nitrates?
4. Which acid reacts to form chlorides?

Carbonic acid is a weak dibasic acid, H_2CO_3.

5. Write an equation showing the ionisation of carbonic acid.
6. What is meant by the term weak acid?
7. What is a dibasic acid?
8. What salts are produced by carbonic acid?

Finish the following word equations.

9. dilute acid + fairly reactive metal → ____________ + ____________
10. dilute acid + metal oxide → ____________ + ____________
11. dilute acid + metal carbonate → ____________ + ____________ + ____________
12. dilute acid + an alkali → ____________ + ____________

1. Ethanoic acid; 2. Sulphuric acid; 3. Nitric acid; 4. Hydrochloric acid;
5. $H_2CO_3 \rightleftharpoons 2H^+ + CO_3^{2-}$; 6. Only partially ionised; 7. Two replaceable hydrogen atoms;
8. Carbonates; 9. salt + hydrogen; 10. salt + water; 11. salt + water + carbon dioxide;
12. salt + water.

Neutralisation

AQA
Edexcel A | Edexcel B
OCR A A
NICCEA
WJEC

KEY POINT **Neutralisation is the reaction of acid and alkali, in the correct proportions, to produce a neutral substance.**

A neutral solution contains equal concentrations of H^+ and OH^- ions. An acidic solution contains an excess of H^+ ions and an alkaline solution an excess of OH^- ions.

Examples of neutralisation

1. Soil that is too acidic does not grow crops as well as more neutral soil. **Slaked lime** (calcium hydroxide) or **limestone** (calcium carbonate) can be added to the soil to neutralise it.
2. **Hydrochloric acid** in the stomach helps in the digestion of food. **Indigestion** is caused by excess acid. The pain can be relieved by taking a **weak alkali** such as sodium hydrogencarbonate or magnesium hydroxide.
3. Insect bites and stings involve an injection of a small amount of chemical into the skin. These chemicals cause irritation. Nettle stings, bee stings and ant bites involve methanoic acid being injected into the skin. Wasp stings involve injecting an alkali into the skin. The irritation can be removed by neutralisation of the acid or alkali.
4. Coal-fired power stations produce sulphur dioxide, which can produce **acid rain**. The sulphur dioxide can be removed from the waste gases before they escape into the atmosphere. Limestone removes sulphur dioxide from the waste gases.
5. Acid rain, caused by sulphur dioxide escaping into the atmosphere, can make lakes and rivers acidic. This can affect organisms in the water such as killing fish. Blocks of limestone put into the water can reduce the acidity.

KEY POINT **Neutralisation can be summarised by the ionic equation:**

$$H^+(aq) + OH^-(aq) \rightarrow H_2O(l)$$

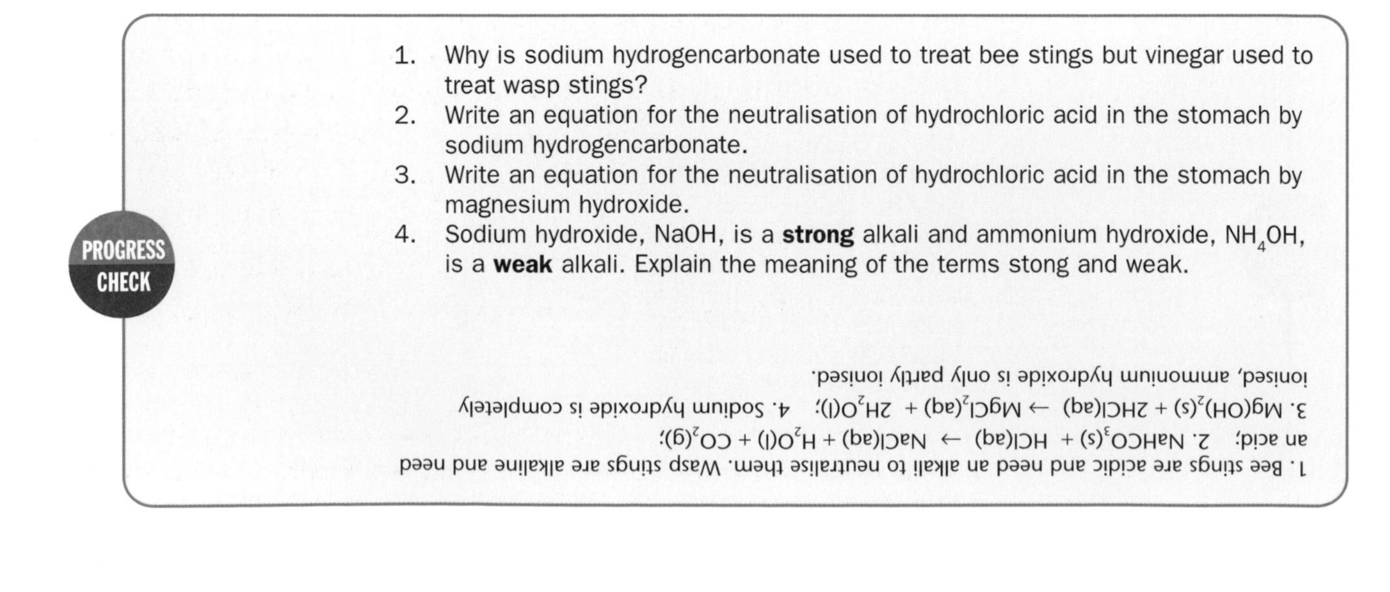

PROGRESS CHECK

1. Why is sodium hydrogencarbonate used to treat bee stings but vinegar used to treat wasp stings?
2. Write an equation for the neutralisation of hydrochloric acid in the stomach by sodium hydrogencarbonate.
3. Write an equation for the neutralisation of hydrochloric acid in the stomach by magnesium hydroxide.
4. Sodium hydroxide, NaOH, is a **strong** alkali and ammonium hydroxide, NH_4OH, is a **weak** alkali. Explain the meaning of the terms stong and weak.

1. Bee stings are acidic and need an alkali to neutralise them. Wasp stings are alkaline and need an acid; 2. $NaHCO_3(s) + HCl(aq) \rightarrow NaCl(aq) + H_2O(l) + CO_2(g)$; 3. $Mg(OH)_2(s) + 2HCl(aq) \rightarrow MgCl_2(aq) + 2H_2O(l)$; 4. Sodium hydroxide is completely ionised, ammonium hydroxide is only partly ionised.

Preparing soluble salts

AQA
Edexcel A Edexcel B
OCR A [A]
NICCEA
WJEC

Soluble salts are salts that readily dissolve in water.

Common soluble salts are:

- all metal nitrates
- all metal chlorides except silver chloride and lead chloride
- all metal sulphates except lead sulphate and barium sulphate. (Calcium sulphate is only sparingly soluble)
- sodium, potassium and ammonium carbonates.

Which method is used to produce a particular salt depends upon various factors – availability, cost, the speed of the reaction – not too fast or not too slow.

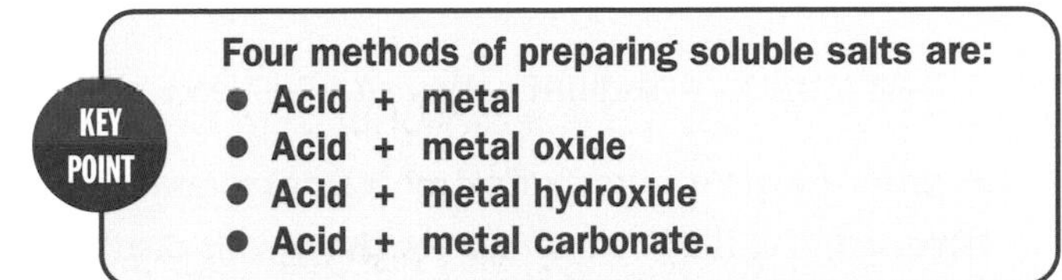
KEY POINT

Four methods of preparing soluble salts are:

- **Acid + metal**
- **Acid + metal oxide**
- **Acid + metal hydroxide**
- **Acid + metal carbonate.**

The method used in each case is the same and is summarised in **Fig. 5.1.**

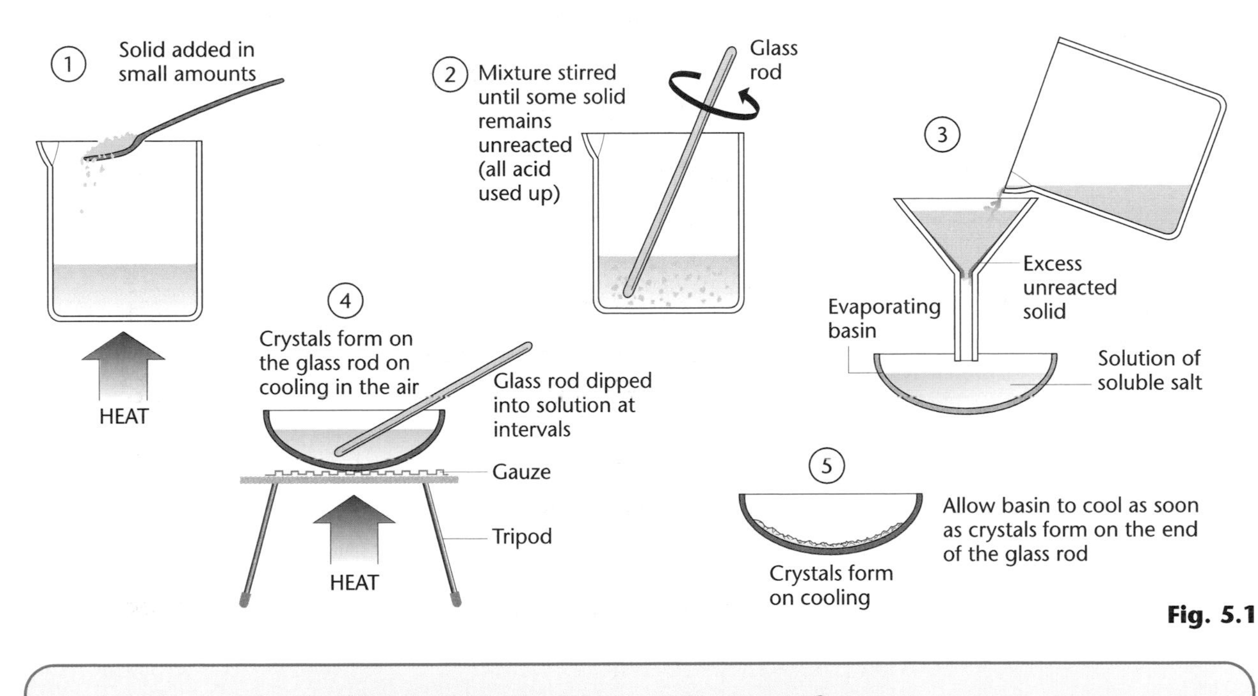

Fig. 5.1

PROGRESS CHECK

1. Which salts in the list are soluble in water?
   sodium sulphate   sodium carbonate   lead carbonate   silver nitrate   barium chloride
2. Magnesium sulphate can be prepared from metal, metal oxide, metal hydroxide or metal carbonate.
   Write equations for these four possible reactions.
3. Magnesium can be extracted from magnesium carbonate. Why is the reaction between magnesium and sulphuric acid unlikely to be used to produce magnesium sulphate on a large scale?

1. sodium sulphate   sodium carbonate   silver nitrate   barium chloride;
2. $Mg(s) + H_2SO_4(aq) \rightarrow MgSO_4(aq) + H_2(g)$; $MgO(s) + H_2SO_4(aq) \rightarrow MgSO_4(aq) + H_2O(l)$; $Mg(OH)_2(s) + H_2SO_4(aq) \rightarrow MgSO_4(aq) + 2H_2O(l)$; $MgCO_3(s) + H_2SO_4(aq) \rightarrow MgSO_4(aq) + H_2O(l) + CO_2(g)$; 3. Extracting magnesium from magnesium sulphate would incur extra costs.

Preparing insoluble salts

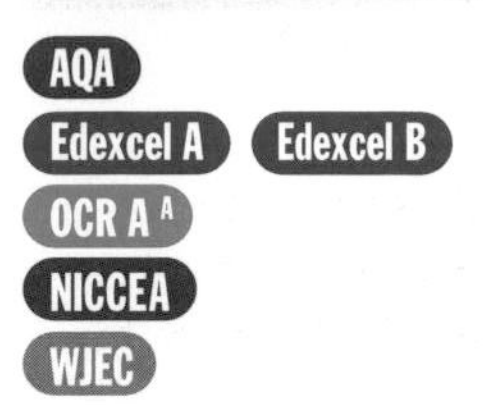

KEY POINT **Insoluble salts are prepared by the process of precipitation.**

This involves mixing two solutions each containing half of the required salt. The salt is then **precipitated**.

Common insoluble salts include silver chloride, barium sulphate, lead sulphate and carbonates, except those of sodium, potassium and ammonium.

e.g. Lead carbonate can be prepared by mixing together solutions of a lead salt (lead nitrate) and a soluble carbonate (sodium carbonate).

$$Pb(NO_3)_2(aq) + Na_2CO_3(aq) \rightarrow PbCO_3(s) + 2NaNO_3(aq)$$

lead nitrate + sodium carbonate → lead carbonate + sodium nitrate

In order to get a pure sample of lead carbonate the mixture should be **filtered**, the lead carbonate **washed** with distilled water and **dried**.

KEY POINT **This type of reaction is sometimes called double decomposition and is represented by the equation:**
AX + BY → AY + BX

PROGRESS CHECK

1. Pick two substances from the list that could be used to prepare lead sulphate by precipitation:
   lead nitrate **lead carbonate** **sodium sulphate** **sodium carbonate**
2. Write an ionic equation for the reaction producing lead carbonate.

Write down the name of the insoluble salt precipitated when each of the following pairs of solutions are mixed:

3. Sodium sulphate and barium chloride
4. Silver nitrate and sodium bromide
5. Sulphuric acid and lead ethanoate
6. Magnesium sulphate and sodium carbonate

1. Lead nitrate and sodium sulphate; 2. $Pb^{2+} + CO_3^{2-} \rightarrow PbCO_3$; 3. Barium sulphate;
4. Silver bromide; 5. Lead sulphate; 6. Magnesium carbonate.

Testing for ions

Testing for metal ions

Testing for ions present is called qualitative analysis.

The presence of many metal ions can be detected by **flame tests**. When compounds of the metal are heated in a hot, blue Bunsen flame they give the flame a characteristic colour.

The compound being tested is mixed with a little concentrated hydrochloric acid.

A piece of clean platinum (or Nichrome) wire is dipped into the mixture and then held in the flame. **Fig. 5.2** below shows the orange colour that sodium compounds give the flame.

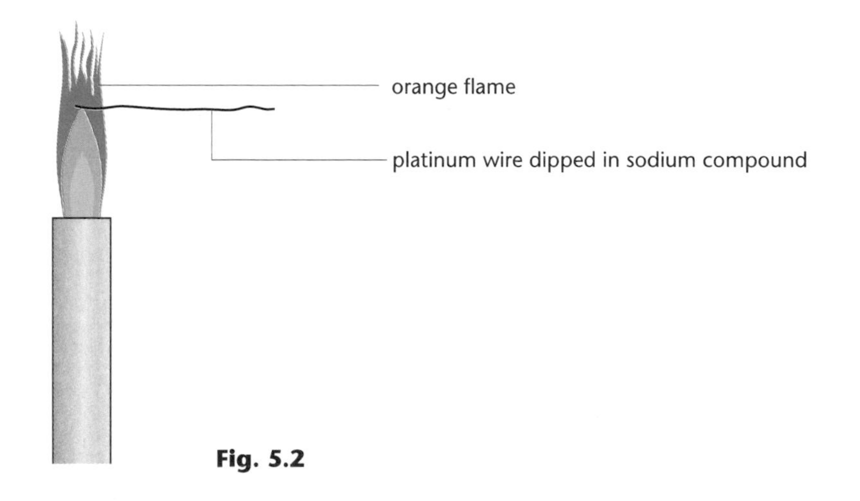

Fig. 5.2

Table 5.2 below shows the characteristic colours of some metal ions.

It is not possible to identify many metal ions, e.g. magnesium, as they give the flame no colour.

Flame test not required for OCR A[A].

Metal ion present	Colour
Sodium	Orange-yellow
Potassium	Lilac-pink
Lithium	Red
Calcium	Brick red
Copper	Green
Lead	Blue
Barium	Pale green

Testing for the ammonium ion

A common mistake is to write the ammonium ion as NH_3^+.

The ammonium ion, NH_4^+, can be tested for by using sodium hydroxide solution.

Sodium hydroxide solution is added to the suspected ammonium compound and the mixture is heated in a test tube.

If an ammonium compound is present, **ammonia** gas is produced. This has a pungent smell and turns damp **red litmus paper** held in the mouth of the test tube **blue** (**Fig. 5.3** see below).

Sodium hydroxide turns red litmus blue so it is important that the red litmus paper is held in the gas and not touching the side of the test tube where sodium hydroxide might have been.

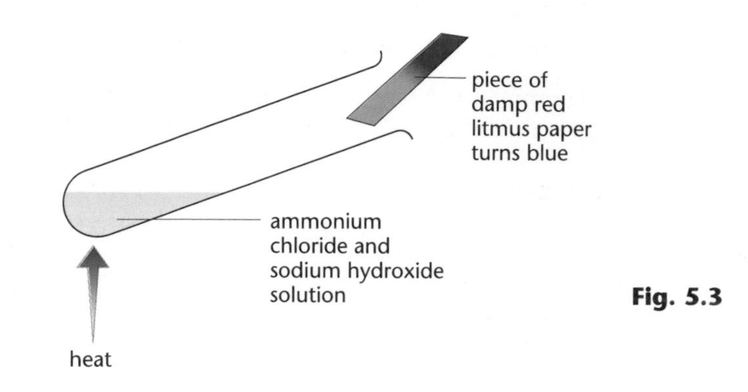

Fig. 5.3

Testing with sodium hydroxide solution

If a small quantity of the compound in solution is tested with **sodium hydroxide solution** an **insoluble hydroxide** may be precipitated. If a precipitate is formed it may redissolve in excess sodium hydroxide solution.

A summary of the precipitation of metal hydroxides with sodium hydroxide solution is shown in **Table 5.3** below.

Detection of magnesium ions is only required by AQA.

Calcium, magnesium and aluminium ions form white precipitates but only aluminium hydroxide redissolves.

Metal ion	Addition of sodium hydroxide solution	
	A couple of drops	Excess
Calcium Ca^{2+}	White precipitate	Precipitate insoluble
Magnesium Mg^{2+}	White precipitate	Precipitate insoluble
Aluminium Al^{3+}	White precipitate	Precipitate soluble – colourless solution
Iron(II) Fe^{2+}	Green precipitate	Precipitate insoluble
Iron(III) Fe^{3+}	Red-brown precipitate	Precipitate insoluble
Copper(II) Cu^{2+}	Blue precipitate	Precipitate insoluble

Testing for negative ions (anions) in solution

Carbonate

When dilute **hydrochloric acid** is added to a carbonate, **carbon dioxide** gas is produced. No heat is required. The carbon dioxide turns limewater milky.

Sulphate

When dilute hydrochloric acid and **barium chloride** solution are added to a solution of a sulphate, a white precipitate of **barium sulphate** is formed immediately.

Chloride, bromide and iodide

When a solution of chloride, bromide or iodide is acidifed with dilute nitric acid and **silver nitrate** added, a precipitate is formed. A chloride produces a white precipitate of **silver chloride**, a bromide a cream precipitate of **silver bromide** and an iodide a yellow precipitate of **silver iodide**.

Test for sulphite ion is required only for EDEXCEL A and B.

Sulphite

When dilute **hydrochloric acid** is added to a sulphite and the mixture is heated, colourless **sulphur dioxide** is formed. Sulphur dioxide turns a piece of filter paper soaked in potassium dichromate(VI) green.

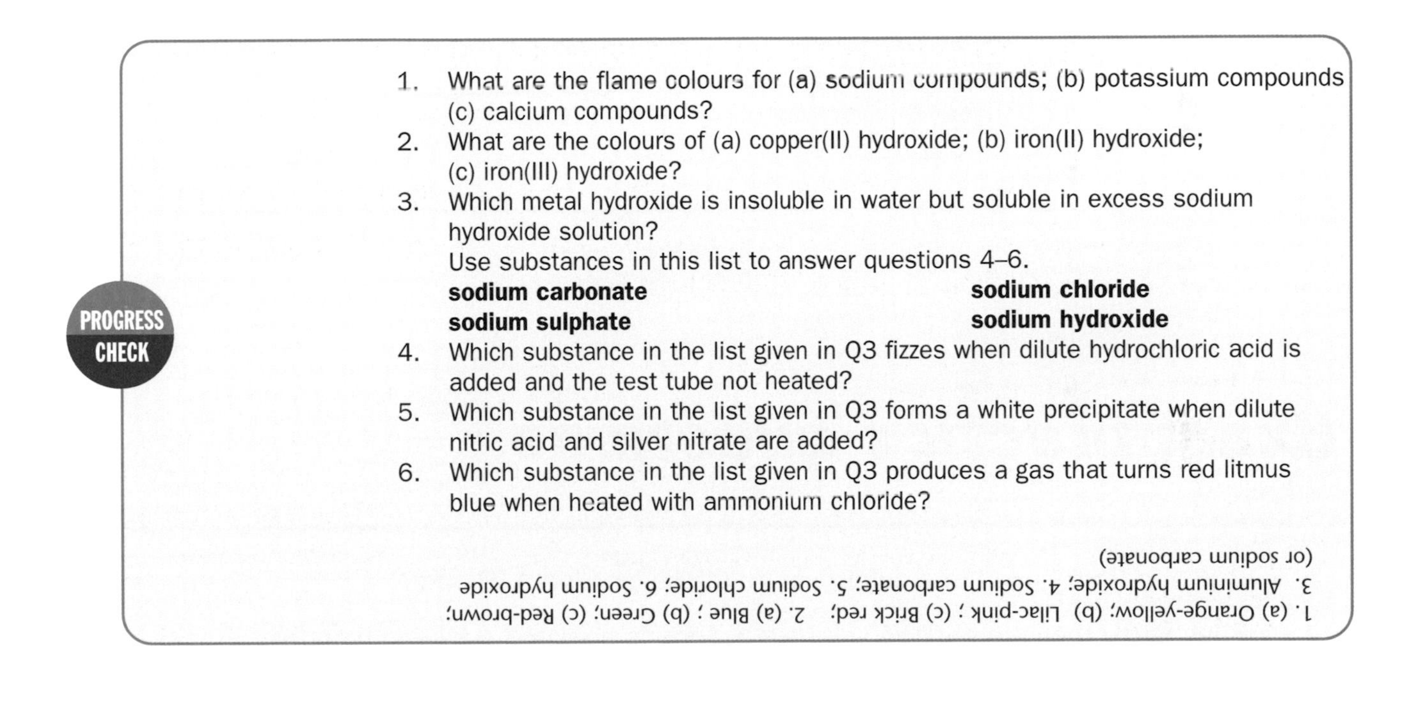

PROGRESS CHECK

1. What are the flame colours for (a) sodium compounds; (b) potassium compounds (c) calcium compounds?
2. What are the colours of (a) copper(II) hydroxide; (b) iron(II) hydroxide; (c) iron(III) hydroxide?
3. Which metal hydroxide is insoluble in water but soluble in excess sodium hydroxide solution?

Use substances in this list to answer questions 4–6.

sodium carbonate **sodium chloride**
sodium sulphate **sodium hydroxide**

4. Which substance in the list given in Q3 fizzes when dilute hydrochloric acid is added and the test tube not heated?
5. Which substance in the list given in Q3 forms a white precipitate when dilute nitric acid and silver nitrate are added?
6. Which substance in the list given in Q3 produces a gas that turns red litmus blue when heated with ammonium chloride?

1. (a) Orange-yellow; (b) Lilac-pink ; (c) Brick red; 2. (a) Blue ; (b) Green; (c) Red-brown; 3. Aluminium hydroxide; 4. Sodium carbonate; 5. Sodium chloride; 6. Sodium hydroxide (or sodium carbonate)

Sample GCSE question

1. Copper(II) sulphate crystals, $CuSO_4.5H_2O$, can be made by adding excess copper(II) oxide, CuO, which is insoluble in water, to dilute sulphuric acid. The mixture is heated.

(a) Why is it necessary to use excess copper(II) oxide? **[1]**

To make sure all of the acid has reacted ✓.

A common mistake here is to state that solution is saturated because copper(II) oxide remains.

(b) How is the excess copper(II) oxide removed? **[1]**

By filtering ✓.

(c) Describe how copper(II) sulphate crystals are formed from this solution. **[3]**

Evaporate the solution ✓. Until a small volume remains ✓. Allow to cool and crystallise ✓.

It is important that the solution is not evaporated until all the water has gone (evaporated to dryness). Under these conditions crystals would not form and anhydrous copper(II) sulphate would remain.

Exam practice questions

1. Sue and Sam have three colourless solutions labelled A, B and C.

They know they are sodium sulphate, sulphuric acid and sodium chloride,

(a) Suggest a series of tests they could do to find out which is which. **[6]**

(b) Write an ionic equation for one reaction they used. **[2]**

2. Lead chromate(VI) is an insoluble salt. It is used as a yellow paint pigment.

Describe how a solution of potassium chromate (VI), K_2CrO_4, could be used to produce a dry sample of lead chromate(VI). Write a balanced symbol equation for the reaction. **[8+1]**

1 mark for Quality of Written Communication

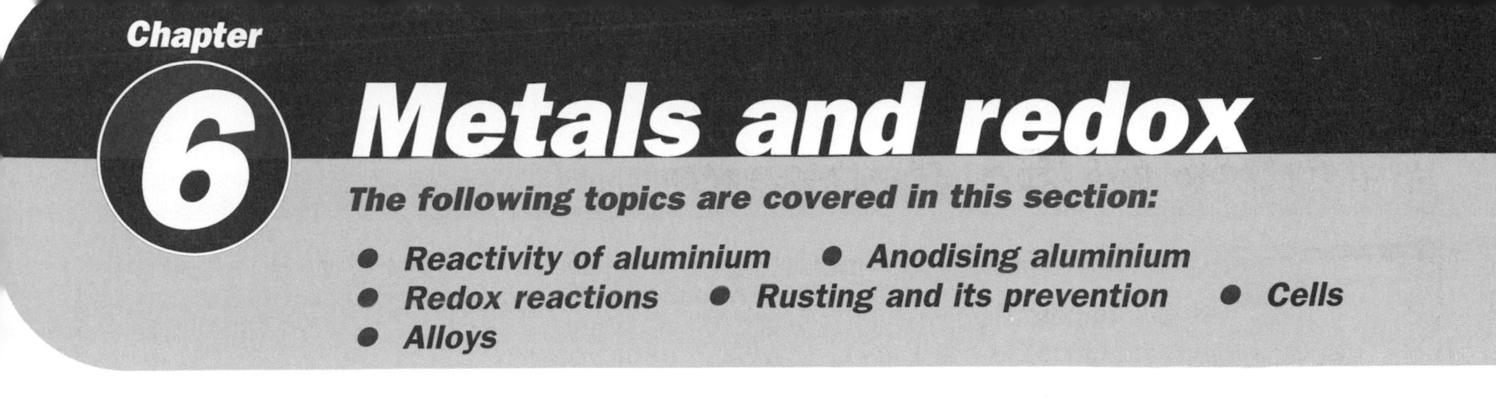

Chapter 6 Metals and redox

The following topics are covered in this section:

- *Reactivity of aluminium*
- *Anodising aluminium*
- *Redox reactions*
- *Rusting and its prevention*
- *Cells*
- *Alloys*

LEARNING SUMMARY

After studying this section you should be able to:

- ***explain why aluminium seems less reactive than would be expected from its position in the reactivity series***
- ***explain oxidation in terms of electron transfer and recall the term redox reaction***
- ***recall that rusting of iron and steel requires both air and water to be present and describe ways of preventing rusting***
- ***describe how a fuel cell can be made and the advantages of fuel cells***
- ***recall the names, composition and uses of alloys.***

Metals are used in a wide range of applications. Iron is the metal used in the largest amounts and most of this is in the form of steel. Although steel is useful it rusts in contact with air and water. Steel is an alloy and alloys are used more than pure metals.

Reactivity of aluminium

AQA
Edexcel A Edexcel B
WJEC

Aluminium is below magnesium but above zinc in the reactivity series.

It would be expected that aluminium is less reactive than magnesium but more reactive than zinc.

Table 6.1 below shows this is not so.

Potassium
Sodium
Calcium
Magnesium
Aluminium
Zinc
Iron

Metal	**Reaction with dilute hydrochloric acid**
Magnesium	Forms bubbles of gas steadily
Aluminium	No reaction after 10 minutes
Zinc	Slow reaction

Aluminium reacts much less than would be expected from its position in the reactivity series. The surface of aluminium is coated with a thin layer of aluminium oxide. Aluminium oxide is very unreactive. This unreactive aluminium oxide layer keeps reactants away from the aluminium and so prevents reaction.

Aluminium has a dull appearance. This is due to the oxide layer. If this layer is removed aluminium becomes more reactive. A piece of aluminium foil heated in a Bunsen flame does not burn. Dipping aluminium foil into mercury removes the oxide layer. Aluminium foil then reacts on standing in the air to form aluminium oxide.

Anodising aluminium

AQA
Edexcel A Edexcel B
WJEC

As previously stated, aluminium is coated with a thin layer of aluminium oxide.

KEY POINT
The process of anodising is used to thicken the layer by electrolysis to give further protection.

This oxide layer can also be coloured by dyes absorbed into the layer to give a decorative finish.

Fig. 6.1 below shows apparatus suitable for anodising a piece of aluminium. Aluminium is used as the anode of the cell. Dilute sulphuric acid is the electrolyte. During electrolysis, oxygen is produced at the anode and this reacts with the aluminium.

$2O^{2-} \rightarrow O_2 + 4e^-$

$4Al(s) + 3O_2(g) \rightarrow 2Al_2O_3(s)$ (overall).

Remember that anodising takes place at the anode so the piece of aluminium is made the anode.

Anodised aluminium is used for making window frames, racing yachts and windsurfing boards.

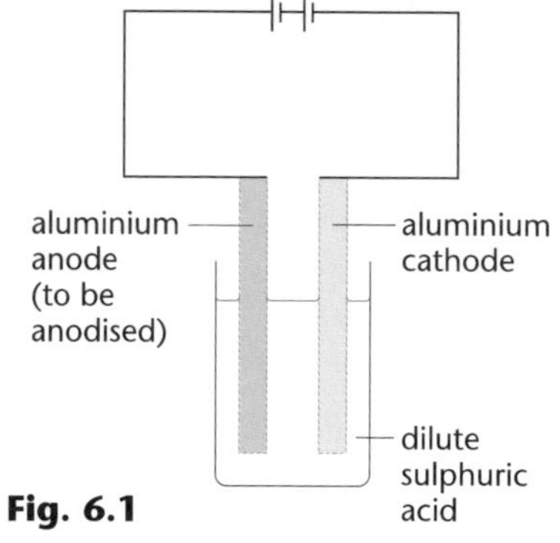

Fig. 6.1

PROGRESS CHECK

1. What is formed as a thin layer on the surface of aluminium?
2. Why does this make the aluminium less reactive than expected?
3. Aluminium placed in dilute sulphuric acid does not react for a long period of time. Eventually after half-an-hour the reaction starts and bubbles of gas are collected. Explain why there is this delay.
4. During anodising, which ion migrates towards the cathode?
5. What is produced at the cathode?
6. Write an ionic equation for the discharging of the ions at the cathode.

1. Aluminium oxide; 2. It prevents the reactant coming into contact with the aluminium surface; 3. It takes a long time for the aluminium oxide surface coating to break down and then reaction can take place; 4. Hydrogen ions, H^+; 5. Hydrogen; 6. $2H^+ + 2e^- \rightarrow H_2$

Redox reactions

OCR A
NICCEA

KEY POINT
Oxidation and reduction reactions occur together. If one substance is oxidised another substance is reduced.
A reaction where oxidation and reduction are taking place is called a redox reaction.

Example: if a mixture of lead(II) oxide and carbon are heated together, the following reaction takes place:

$PbO(s) + C(s) \rightarrow Pb(s) + CO(g)$

lead(II) oxide + carbon → lead + carbon monoxide

You should know that a reaction where oxygen is gained or hydrogen is lost is called an oxidation reaction. A reaction where oxygen is lost or hydrogen is gained is called a reduction reaction.

In this reaction lead(II) oxide is losing oxygen and carbon is gaining oxygen. Lead oxide is **reduced** and carbon is **oxidised**. Reduction and oxidation are taking place and this is called a **redox reaction**.

No reaction would take place if the lead(II) oxide was heated alone. Carbon is the substance which is necessary for the reduction to take place because it removes the oxygen. Carbon is called the **reducing agent**.

A reducing agent is a substance that reduces some other substances but is itself oxidised.

Similarly, lead(II) oxide is the **oxidising agent**. It supplies oxygen, which is used to oxidise the carbon.

An oxidising agent is a substance that oxidises some other substances but is itself reduced.

Common reducing agents include hydrogen, carbon and carbon monoxide.

Common oxidising agents include oxygen, chlorine, concentrated sulphuric acid and concentrated nitric acid.

Oxidation and reduction in terms of electron transfer

A more advanced definition of oxidation and reduction can be made in terms of loss and gain of electrons.

A simple mnemonic to remember. OILRIG oxidation is loss and reduction is gain (of electrons).

Oxidation is any process where electrons are lost and reduction is any process where electrons are gained.

Common examples are:

1. Iron(II) to iron(III) ions

$$Fe^{2+} \rightarrow Fe^{3+} + e^-$$

Each iron(II) ion loses one electron to become an iron(III) ion. This is **oxidation** (loss of electrons)

2. Chlorine to chloride ions

$$Cl_2 + 2e^- \rightarrow 2Cl^-$$

Chlorine molecules gain electrons to become chloride ions. This is **reduction** (gain of electrons).

Other halogens, bromine and iodine, act in a similar way.

$$Br_2 + 2e^- \rightarrow 2Br^-$$

$$I_2 + 2e^- \rightarrow 2I^-$$

Reaction of chlorine and iron(II) chloride:

If chlorine is bubbled through iron(II) chloride solution, iron(II) ions are oxidised to iron(III) ions and chlorine is reduced to chloride ions.

$$2FeCl_2(aq) + Cl_2(g) \rightarrow 2FeCl_3(aq)$$

$$Fe^{2+} \rightarrow Fe^{3+} + e^-$$

$$Cl_2 + 2e^- \rightarrow 2Cl^-$$

Multiply equation 1 by 2 and add the two equations together.

$$2Fe^{2+} + Cl_2 \rightarrow 2Fe^{3+} + 2Cl^-$$

These are called half-equations. Two half equations can be added together to form an ionic equation.

PROGRESS CHECK

1. In the reaction of iron(II) ions and chlorine, which substance is the oxidising agent and which is the reducing agent?

When chlorine is bubbled through potassium iodide solution a displacement reaction takes place and iodine and chloride ions are produced.

2. Write an ionic equation for this reaction.
3. Which substance is oxidised and which substance is reduced? Explain your answer.

When a piece of zinc is dipped into copper(II) sulphate solution a displacement reaction takes place.

4. Write an equation for the reaction taking place.
5. Write an ionic equation for the reaction.
6. Write two ionic half equations for the changes taking place.
7. Explain what is happening in the reaction in terms of oxidation and reduction.

1. Iron(II) ions are the reducing agent (they are oxidised) and chlorine is the oxidising agent (it is reduced); 2. $Cl_2 + 2I^- \rightarrow 2Cl^- + I_2$; 3. Chlorine is reduced (gains electrons) and iodide ions are oxidised (lose electrons); 4. $Zn + CuSO_4 \rightarrow ZnSO_4 + Cu$; 5. $Zn + Cu^{2+} \rightarrow Zn^{2+} + Cu$; 6. $Zn \rightarrow Zn^{2+} + 2e^-$, $Cu^{2+} + 2e^- \rightarrow Cu$; 7. Zinc is oxidised (loses electrons), copper(II) ions are reduced (gain electrons)

Rusting and its prevention

OCR A [A]
NICCEA
WJEC

Iron and steel react in the atmosphere to produce reddish-brown rust.

The chemical composition of rust is complicated. It is best regarded as a hydrated iron(III) oxide, $Fe_2O_3.xH_2O$. The rusting process is an **oxidation** reaction and can be represented by the equation

$$Fe \rightarrow Fe^{3+} + 3e^-$$

A surface coating of aluminium oxide forms on a piece of aluminium and prevents further reaction. Rust, however, flakes off and shows a fresh metal surface for rusting.

Fig. 6.2 below shows an experiment used to find the conditions needed for rusting.

Rusting in the United Kingdom costs an estimated £2 500 000 000 each year.

Fig. 6.2

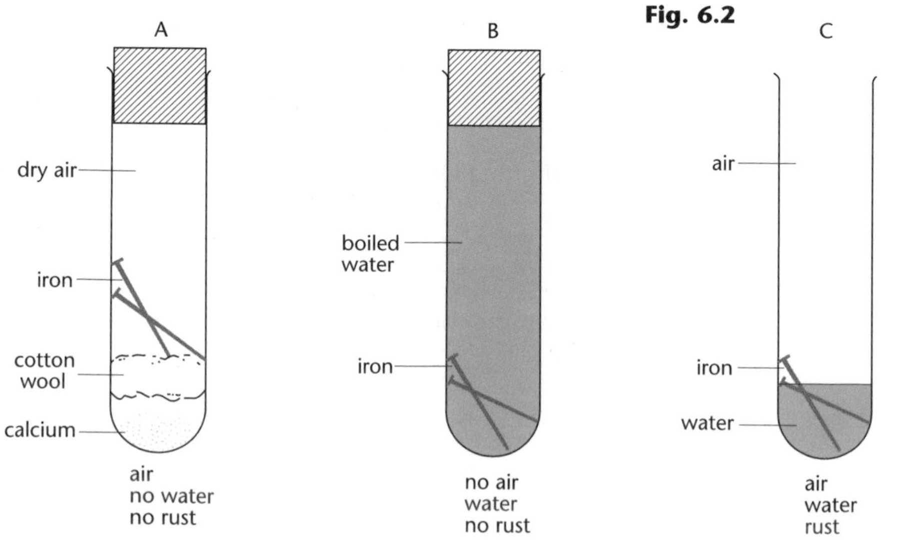

If the three test tubes are set up as shown in **Fig. 6.2** and left for a few days, rusting only occurs in test tube C.

From this it can be concluded that:

Both air and water are needed for rusting to take place.

In fact, it is the **oxygen** in the air along with the **water** that are needed for rusting.

Rusting is speeded up when acid (carbon dioxide or sulphur dioxide) or salt are present.

Ways of preventing rusting

Rusting can be prevented if air (or oxygen) and/or water can be excluded from the iron or steel.

1. **Painting**. A coat of paint prevents oxygen and water coming into contact with the iron or steel. This will prevent rusting only while the coating is not broken. This type of rust prevention is used to prevent steel car bodies or iron railings rusting.
2. **Oil or grease**. These prevent oxygen and water coming in contact with iron and steel. They are useful when parts are moving.
3. **Coating with zinc**. A layer of zinc on iron or steel prevents rusting by stopping oxygen and water coming into contact with the metal. This is called **galvanising**. It is used for fences. It cannot be used for food cans as zinc compounds are poisonous. Tin plate is used to make food cans.
4. **Sacrificial protection**. This is used for preventing the rusting of steel hulls of ships or steel legs of piers. To prevent a steel hull rusting, blocks of **magnesium** or **zinc** are strapped to the steel hull. The metal used must be higher in the reactivity series. The magnesium or zinc blocks corrode in preference to iron. As long as they remain no rusting will take place. These blocks can easily be replaced.
5. **Electroplating**. A steel item can be protected from rusting by electroplating. A thin layer of **nickel** is plated on the surface to prevent oxygen and water coming in contact with the steel. A very thin coating of **chromium** is often plated on top to give a decorative appearance.

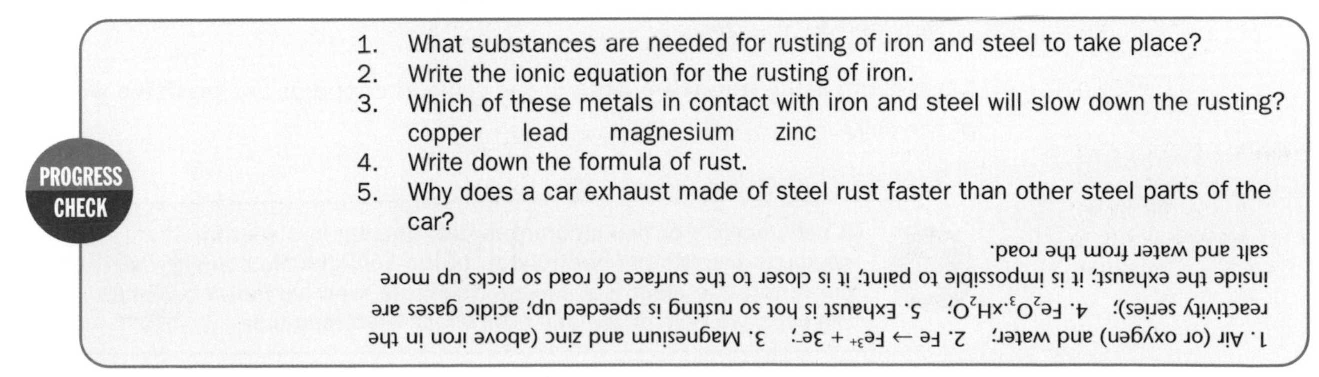
PROGRESS CHECK

1. What substances are needed for rusting of iron and steel to take place?
2. Write the ionic equation for the rusting of iron.
3. Which of these metals in contact with iron and steel will slow down the rusting?
   copper lead magnesium zinc
4. Write down the formula of rust.
5. Why does a car exhaust made of steel rust faster than other steel parts of the car?

1. Air (or oxygen) and water; 2. $Fe \rightarrow Fe^{3+} + 3e^-$; 3. Magnesium and zinc (above iron in the reactivity series); 4. $Fe_2O_3.xH_2O$; 5. Exhaust is hot so rusting is speeded up; acidic gases are inside the exhaust; it is impossible to paint; it is closer to the surface of road so picks up more salt and water from the road.

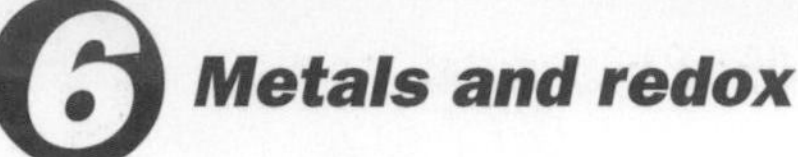

Cells

OCR A B

KEY POINT **Electricity can be produced by cells where a chemical reaction is taking place.**

Fig. 6.3 below shows two pieces of copper dipping in an **electrolyte** (solution of sodium chloride).

The two pieces of copper are connected together with a copper wire and light bulb.

Fig. 6.3

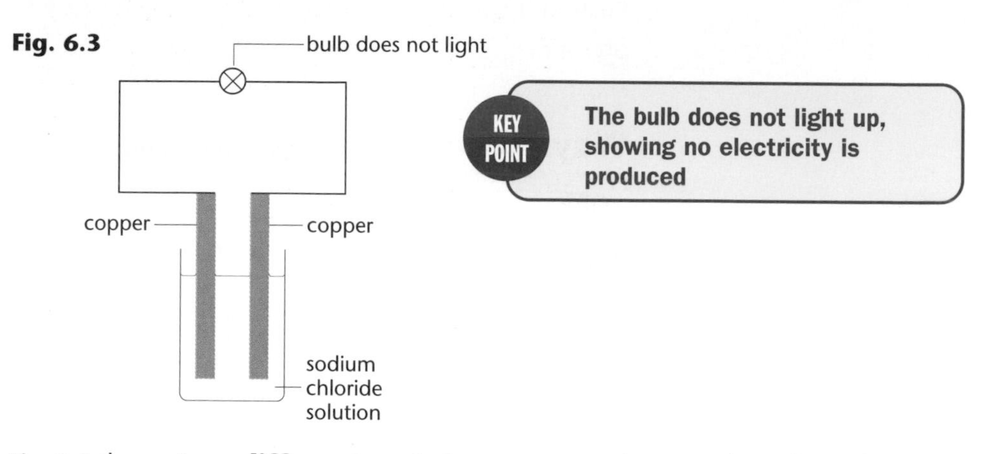

KEY POINT **The bulb does not light up, showing no electricity is produced**

Fig.6.4 shows **two different metals** – copper and magnesium dipped in sodium chloride solution.

In any cell a chemical reaction is taking place. The energy is released as electrical energy. The cell stops working when one of the chemicals is used up.

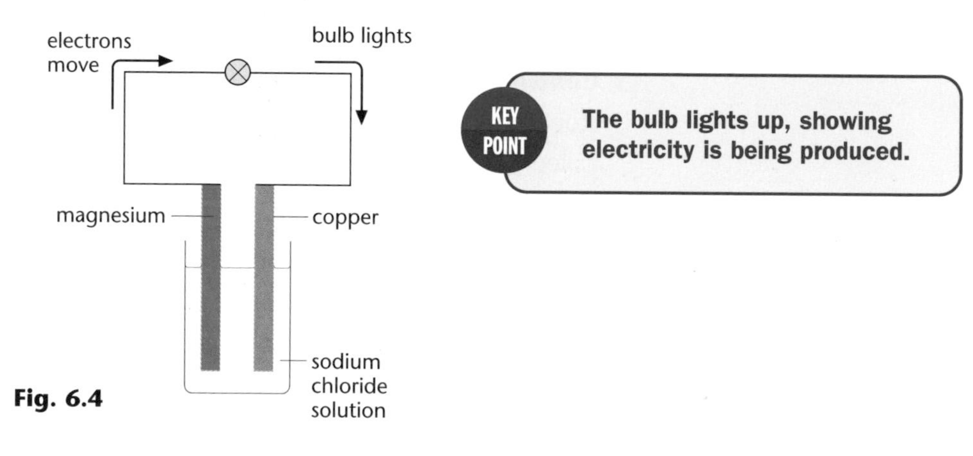

Fig. 6.4

KEY POINT **The bulb lights up, showing electricity is being produced.**

Electrons are moving through the wire from the magnesium rod to the copper rod. If a voltmeter is put in the wire instead of light bulb a voltage of 2.7 volts can be measured. This is a measure of the pushing power in the circuit.

At the magnesium , electrons are lost.

$Mg \rightarrow Mg^{2+} + 2e^-$

Magnesium is the **negative pole** of the cell and copper is the **positive pole** of the cell.

KEY POINT **A cell consists of two different metals dipping in a solution that conducts electricity (electrolyte). In the cell, chemical energy is converted into electrical energy. The more reactive metal becomes the negative pole of the cell from which electrons flow.**

Using different metals in cells

Table 6.2 below shows the voltages in different cells. In each cell, one metal (rod A) is copper and the other metal (rod B) is changed. The same salt solution is used in each case.

Rod A	Rod B	Voltage (V)
Copper	Zinc	0.60
Copper	Iron	0.30
Copper	Lead	0.02
Copper	Copper	0.00
Copper	Silver	–0.5

KEY POINT The further apart the metals are in the reactivity series the higher will be the voltage of the cell.

Fuel cells

KEY POINT A fuel cell is a special type of cell which converts chemical energy into electrical energy in an efficient way.

Fig. 6.5 below shows a diagram of a fuel cell that produces electricity using the reaction between hydrogen and oxygen.

A hydrogen-oxygen fuel cell does not produce any pollution. The only product is water.

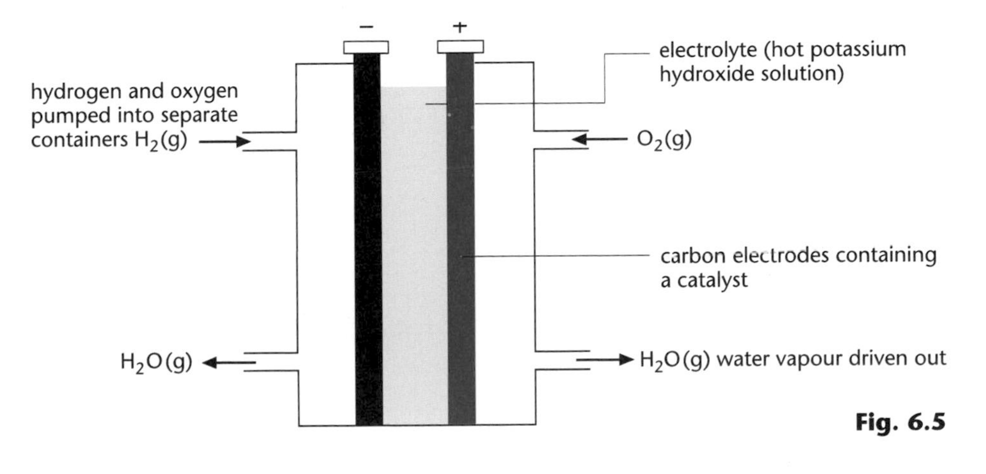

Fig. 6.5

The reactions in the hydrogen–oxygen fuel cell

At the positive electrode:

$$2H_2(g) + 4OH^-(aq) \rightarrow 4H_2O(l) + 4e^-$$

At the negative electrode:

$$O_2(g) + 2H_2O(l) + 4e^- \rightarrow 4OH^-(aq)$$

Overall:

$$2H_2(g) + O_2(g) \rightarrow 2H_2O(l)$$

For fuel cells to be widely used a cheap source of hydrogen is needed.

KEY POINT This is an exothermic reaction and all of the energy produced is produced as electricity.

Fuel cells are used in space vehicles and they are being tried in cars instead of petrol engines.

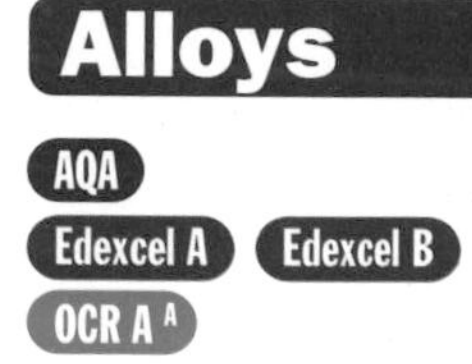

1. A cell is set up with two rods dipping into a solution. Which of the sets of apparatus would produce a voltage on a voltmeter?

	Rod S	Rod T	Solution
A	zinc	lead	glucose
B	zinc	zinc	sodium sulphate
C	zinc	lead	sodium sulphate
D	lead	lead	sodium sulphate

2. A simple cell was set up with copper and silver dipping into sodium chloride solution.
   Which metal is the negative electrode?
3. What is produced in a hydrogen–oxygen fuel cell?
4. Why is an engine burning hydrogen and oxygen better than a petrol-burning engine?
5. Why is a fuel cell better than an engine burning hydrogen?

1. C; 2. Copper (higher in the reactivity series); 3. Water; 4. The only product is water which does not pollute the atmosphere – no carbon dioxide, carbon monoxide or sulphur dioxide; 5. All of the energy is produced as electricity and none is lost as heat to the surroundings.

Alloys

AQA Edexcel A Edexcel B OCR A [A]

Alloys are mixtures of metals or mixtures of metals with carbon. Alloys have many applications because they have better properties for many uses than pure metals.

Table 6.3 below gives the uses of some pure metals.

Metal	Use	Reason for use
Copper	Copper wires	Excellent conductor of electricity/very ductile
Tin coating	tin cans	Not poisonous
Aluminium	Kitchen foil	Very malleable
Iron	Wrought iron gates	Easy to forge
Lead	Flashing on roofs	Soft, easy to shape/very unreactive

An alloy can be made by weighing out correctly the different constituent metals and then melting them together.

Fig. 6.6 shows the arrangement of particles in a pure metal and in an alloy.

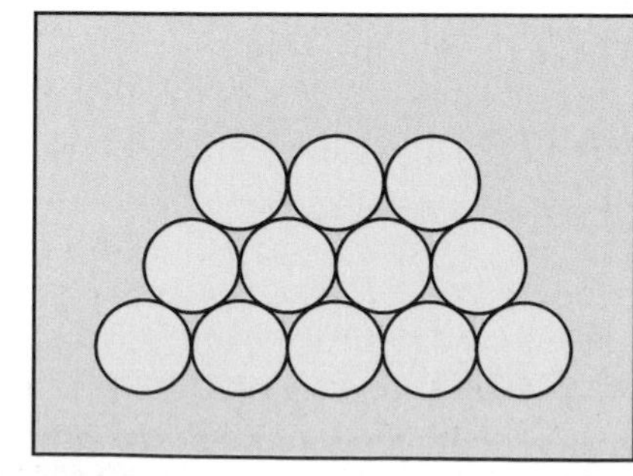

pure metal

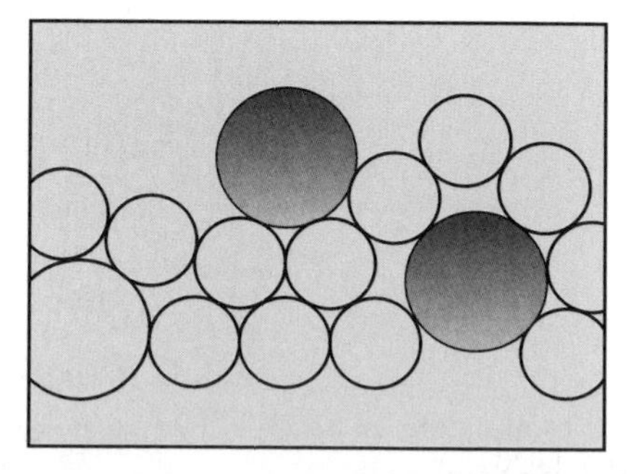

alloy

Fig. 6.6

The particles in a pure metal are able to move past each other much better than the particles in an alloy. This makes the alloy harder.

Table 6.4 below gives the composition and uses of some common alloys.

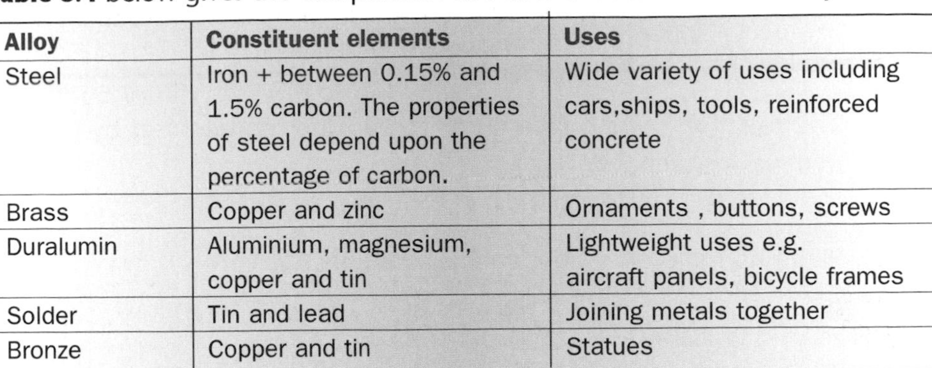

Alloy	Constituent elements	Uses
Steel	Iron + between 0.15% and 1.5% carbon. The properties of steel depend upon the percentage of carbon.	Wide variety of uses including cars,ships, tools, reinforced concrete
Brass	Copper and zinc	Ornaments , buttons, screws
Duralumin	Aluminium, magnesium, copper and tin	Lightweight uses e.g. aircraft panels, bicycle frames
Solder	Tin and lead	Joining metals together
Bronze	Copper and tin	Statues

Steel

Steel is required for AQA and EDEXCEL A & B.

KEY POINT

Most iron produced in a blast furnace is turned into steel. The iron from the blast furnace, called pig iron, contains impurities of carbon, phosphorus and silicon.

The steel-making process involves removing all of the impurities to get pure iron and then adding the correct amounts of different materials to get steel of the required quality.

The steel-making furnace **(Fig. 6.7)** is tilted and loaded with 30% scrap iron and 70% molten iron from the blast furnace.

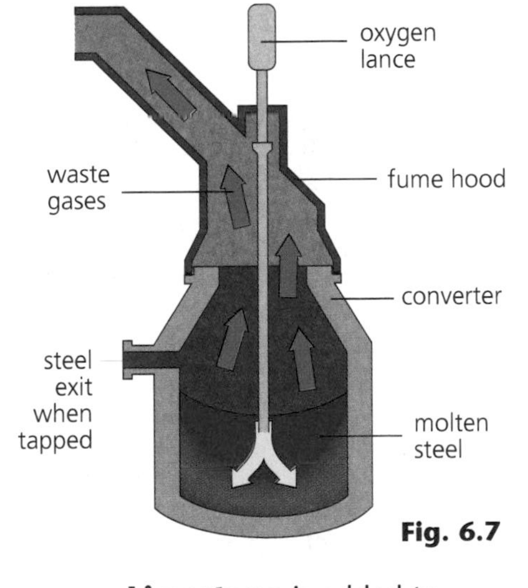

Fig. 6.7

A water-cooled lance is lowered into the upright furnace and pure **oxygen** is blown, under high pressure, onto the surface of the molten iron. The oxides of carbon and phosphorus escape as gases. **Limestone** is added to remove the other impurities as slag. Finally, the required amounts of carbon and other elements are added to give steel of the required quality.

Table 6.5 compares the properties and uses of three types of steel.

Type of steel	Properties of steel	Percentage of carbon
Low carbon steel	Soft and easily shaped	0.03–0.25
Mild steel	Easily pressed into shape	0.25–0.50
High carbon steel	Strong but brittle	0.85–1.50

Stainless steel contains transition metals such as chromium and nickel. It does not rust easily as do iron and other types of steel.

PROGRESS CHECK

1. Which type of steel is most suitable for making drill bits?
2. Why is it better to remove all of the impurities from steel during manufacture and then add the carbon again afterwards?
3. Solder is an alloy of tin and lead. It has a lower melting point than tin or lead. Why is this an advantage when soldering metals together?
4. Why is duralumin better for constructing aeroplanes but pure aluminium better for overhead power cables?

1. High carbon steel; 2. By removing all impurities it is easier to get the correct amounts in the steel at the end of the process; 3. When metals are soldered the solder has to be melted. A low melting point means it is easier to melt the solder; 4. Duralumin is stronger and this is needed for aircraft structure, but pure aluminium is a better conductor of electricity.

Sample GCSE question

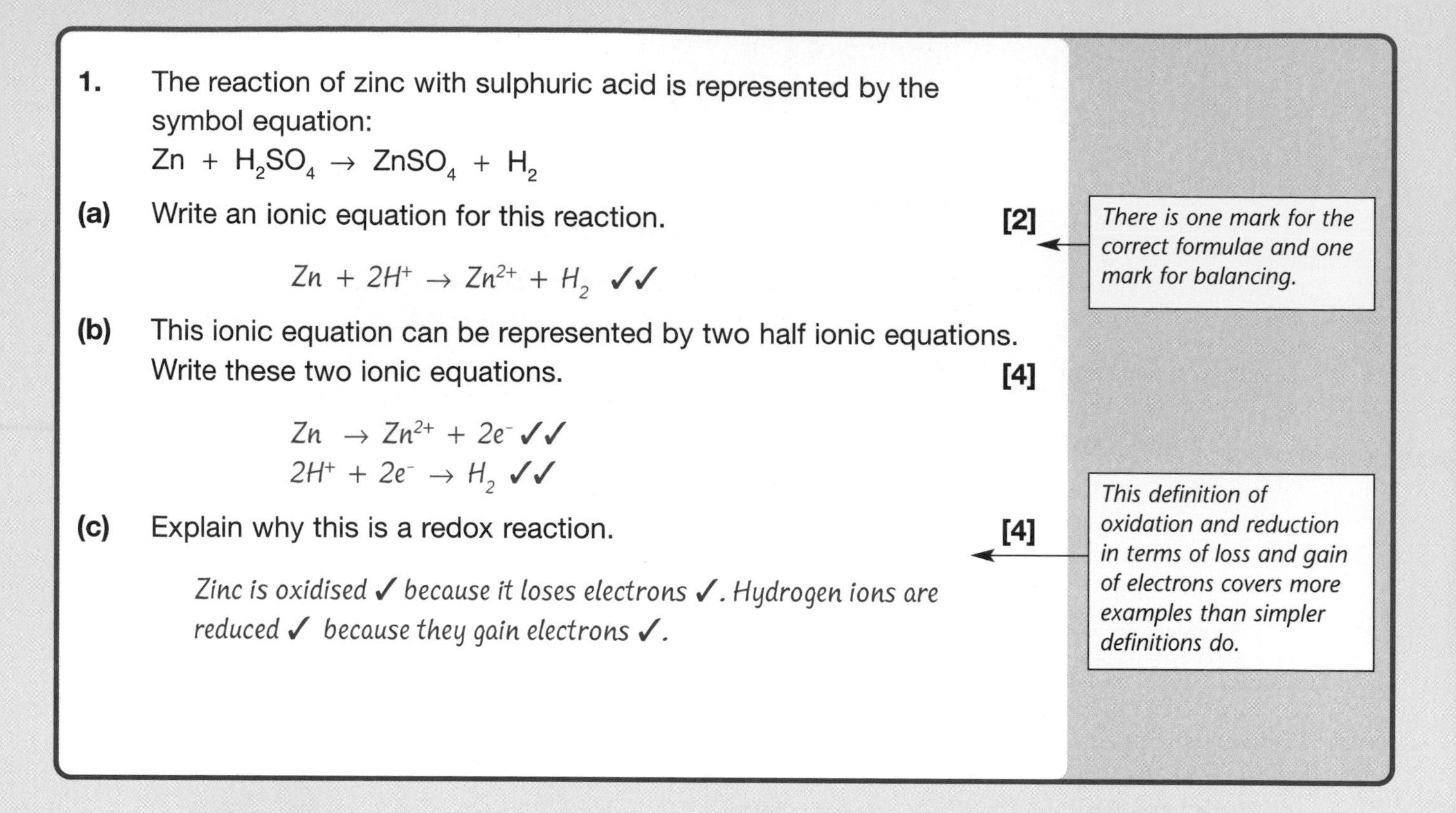

1. The reaction of zinc with sulphuric acid is represented by the symbol equation:

$Zn + H_2SO_4 \rightarrow ZnSO_4 + H_2$

(a) Write an ionic equation for this reaction. **[2]**

$Zn + 2H^+ \rightarrow Zn^{2+} + H_2$ ✓✓

There is one mark for the correct formulae and one mark for balancing.

(b) This ionic equation can be represented by two half ionic equations. Write these two ionic equations. **[4]**

$Zn \rightarrow Zn^{2+} + 2e^-$ ✓✓

$2H^+ + 2e^- \rightarrow H_2$ ✓✓

(c) Explain why this is a redox reaction. **[4]**

Zinc is oxidised ✓ because it loses electrons ✓. Hydrogen ions are reduced ✓ because they gain electrons ✓.

This definition of oxidation and reduction in terms of loss and gain of electrons covers more examples than simpler definitions do.

Exam practice question

1. Steel is the most widely used alloy.

Describe how steel is formed from iron ore. **[7+1]**

This is a question requiring continuous prose in your answer. Seven marks are for Chemistry and one mark is for QWC (Quality of Written Communication). In this case it is for the correct order of the steps. There is one mark for each emboldened statement and one mark for one of the other statements.

Chapter

7 Further carbon chemistry

The following topics are covered in this section:

- ***Hardening natural oils***
- ***Isomerism***
- ***Alcohols***
- ***Ethanoic acid***
- ***Esters***
- ***Further polymers***

LEARNING SUMMARY

After studying this section you should be able to:

- ***recall how natural fats and oils can be hardened by reaction with hydrogen***
- ***explain how different isomers can exist for the same molecular formula***
- ***describe how ethanol can be manufactured by different methods***
- ***discuss the advantages and disadvantages of the two methods for preparing ethanol***
- ***recall that oxidation of ethanol produces ethanoic acid***
- ***describe how esters can be produced.***

Complete combustion and test for alkenes are required for AQA. It will be found in the Core Section (pages 27–28).

KEY POINT

You will know that there are different families of hydrocarbons. Alkanes are saturated hydrocarbons used as fuels and chemical feedstock. Alkanes can be cracked to produce alkenes and alkenes can be polymerised to form addition polymers. In this Unit we will consider other carbon compounds especially those compounds called alcohols.

Hardening natural oils

AQA

OCR A [B]

Alkenes undergo addition reactions (2.1).

If a mixture of **ethene** and **hydrogen** is passed over a heated nickel catalyst, an addition reaction can take place, the product of which is **ethane**.

$$\begin{array}{ccc} H & & H \\ & C{=}C & \\ H & & H \end{array} + H{-}H \rightarrow \begin{array}{c} \quad H \;\; H \\ \quad | \;\;\; | \\ H{-}C{-}C{-}H \\ \quad | \;\;\; | \\ \quad H \;\; H \end{array}$$

Natural oils such as sunflower oil are **liquid** and **unsaturated** i.e. they contain carbon–carbon double bonds.

These oils can be **hardened** by addition reactions with hydrogen. The oil and hydrogen are passed over a nickel catalyst at 170°C.

The resulting fat is used as margarine.

Isomerism

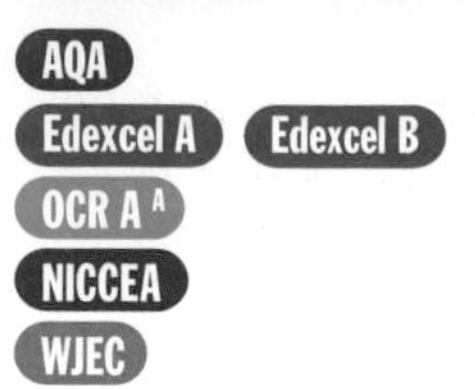

With alkanes containing up to three carbon atoms, there is only one possible structure for each molecular formula.

However, when there are four or more carbon atoms in an alkane it is possible to have different structures.

KEY POINT **Isomerism is the existence of two or more compounds with the same molecular formula but different structural formulae.**

Do not confuse the term isomer with isotope or allotrope.

The two isomers of butane are

These two isomers have similar chemical properties but different physical properties.

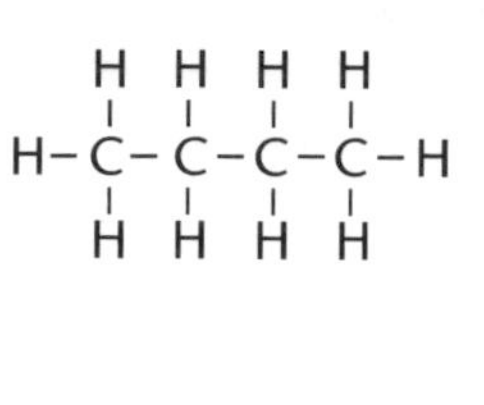

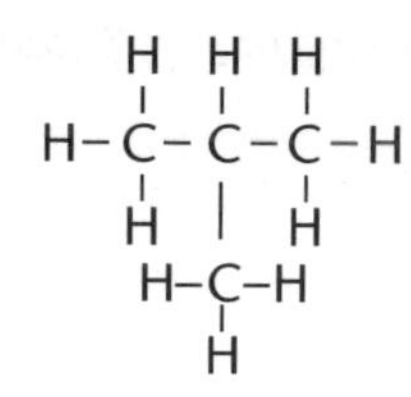

butane

2-methylpropane

In each structure, every carbon atom is joined to 4 other atoms and every hydrogen atom is joined to one other atom.

Isomerism becomes more common with higher alkanes, e.g. there are 75 isomers of decane $C_{10}H_{22}$. Isomers also occur with other homologous series (families), e.g. alkenes, and also with compounds in different homologous series e.g. C_2H_5OH (an alcohol) and CH_3OCH_3 (an ether).

PROGRESS CHECK

1. For each of the following statements write true or false:
(a) Isomers have the same molecular formula.
(b) Isomers have the same relative formula mass.
(c) Isomers have the same number of bonds.
(d) Isomers have the same melting and boiling points.
2. Draw the three isomers of pentane, C_5H_{12}.
3. Butene, C_4H_8 is an alkene and contains a carbon–carbon double bond. Draw the structural formulae of the isomers of butene.

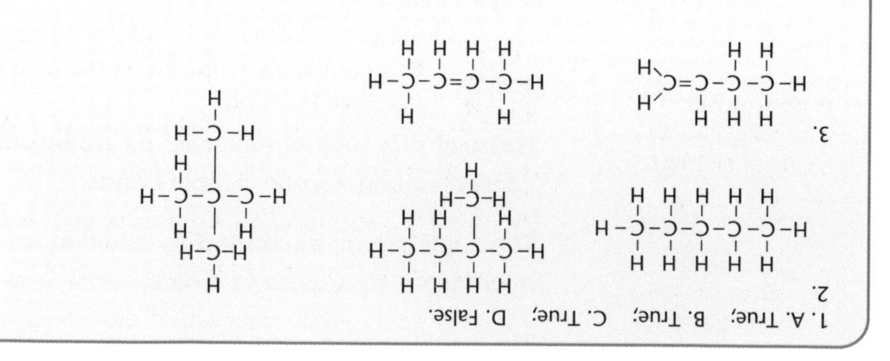

Alcohols

Ethanol C_2H_5OH is an organic chemical of great importance. It is a member of the **alcohol** family. Alcohols have a general formula $C_nH_{2n+1}OH$.

Alcohols are derived from an alkane by removing an –H and adding an –OH.

The structural formula of ethanol is

```
    H  H
    |  |
H – C – C – O – H
    |  |
    H  H
```

Manufacture of ethanol

Large quantities of ethanol are manufactured for industrial use.

There are two methods of producing ethanol – from **ethene** or from **sugar**.

From ethene

Large amounts of ethene are produced from **cracking fractions** from **crude oil**.

Much of this is used to make poly(ethene) but much is used for making ethanol. This involves an **addition reaction**.

Ethene is mixed with steam and passed over a phosphoric acid catalyst at 600°C and at a high pressure.

WJEC does not require the manufacture of ethanol from ethene.

$C_2H_4(g) + H_2O(g) \rightarrow C_2H_5OH(g)$

ethene + steam → ethanol

```
H     H      H   H          H  H
 \   /        \ /           |  |
  C=C    +     O    →   H – C – C – O – H
 /   \                      |  |
H     H                     H  H
```

From sugar

Ethanol can be prepared by the fermentation of sugar solutions using enzymes in yeast.

A dilute solution of ethanol is present in beer and wine.

The solution is kept in a warm place for several days.

The solution produced is a dilute solution of ethanol.

$C_6H_{12}O_6(aq) \rightarrow 2C_2H_5OH(aq) + 2CO_2(g)$

glucose → ethanol + carbon dioxide

Spirits such as whisky and gin are produced by fractional distillation.

A more concentrated solution of ethanol is produced by **fractional distillation**.

There are good opportunities here to compare the manufacture of ethanol by two different methods. There is no better method. It depends upon circumstances.

The method used to manufacture ethanol depends upon the materials available.

1. In developed countries, such as the United States, and in Europe there are large amounts of ethene available. Making ethanol from ethene would be preferred.

2. In countries such as Mauritius, which do not have crude oil but do have sugar produced from sugar cane in large amounts, fermentation would be preferred.

Ethanol made by the fermentation process is a **batch** process. Ethanol is produced from ethene by a **continuous** process.

Table 7.1 below compares the advantages and disadvantages of each method.

Ethanol by fermentation	Ethanol from ethene
Advantages	**Advantages**
Uses **renewable resources** e.g. sugar cane Uses otherwise waste materials	**Fast** reaction, **Continuous** process Does not need large reaction vessels Produces pure ethanol
Disadvantages	**Disadvantages**
Large volume needed to produce small amount of ethanol Needs large reaction vessels Fractional distillation is expensive Fermentation is **slow** When ethanol reaches a certain concentration, the reaction stops	Uses a **non-renewable resource** Energy is needed to produce steam High percentage of ethene remains unreacted and must be re-cycled

Uses of ethanol

When perfume is sprayed on the skin the cooling felt is due to the evaporation of ethanol.

1. Ethanol is widely used as a **solvent**. Ethanol is used for paints, varnishes, perfumes etc. When the product is applied, the ethanol evaporates.

2. Ethanol is used as a **fuel**. In Brazil, either pure ethanol or a mixture of petrol and ethanol is used for fuel in cars. Ethanol burns in excess air to form carbon dioxide and water.

$$C_2H_5OH + 3O_2 \rightarrow 2CO_2 + 3H_2O$$

3. Ethanol is used for **making other organic chemicals** e.g. ethanoic acid, esters.

4. Ethanol is used in **alcoholic drinks**.

Different drinks contain different percentages of alcohol (**see Table 7.2** below)

Drink	Approximate percentage of ethanol
Beers	4
Wine	12
Fortified wine e.g. sherry	18
Spirits (whisky)	35

There is evidence about the harmful effects of ethanol including:

- impaired co-ordination and judgement
- slower reaction times
- promotes aggression
- causes depression and other mental disorders
- causes ulcers, high blood pressure, brain and liver damage.

Pure ethanol cannot be purchased in shops.

We can buy **methylated spirits**. This is ethanol with added **methanol**. Methanol is highly toxic. Other substances are added to make it undrinkable and a purple dye is added as a warning.

PROGRESS CHECK

1. Write the molecular formula of methanol.
2. Draw the structural formula of methanol.
3. Write an equation for the combustion of methanol in excess air.
4. Draw the structural formulae of two alcohols with a molecular formula C_3H_8O.

Ethene, C_2H_4, is produced when ethanol vapour is passed over heated aluminium oxide.

5. Write the equation for this reaction.
6. What type of reaction is taking place?

1. CH_4O or CH_3OH

2.
```
    H
    |
H - C - O - H
    |
    H
```

3. $2CH_3OH + 3O_2 \rightarrow 2CO_2 + 4H_2O$;

4.
```
    H   H   H                   H   H   H
    |   |   |                   |   |   |
H - C - C - C - O - H       H - C - C - C - H
    |   |   |                   |   |   |
    H   H   H                   H   O   H
                                    |
                                    H
```

5. $C_2H_5OH \rightarrow C_2H_4 + H_2O$;

6. Removal of water (dehydration).

Ethanoic acid

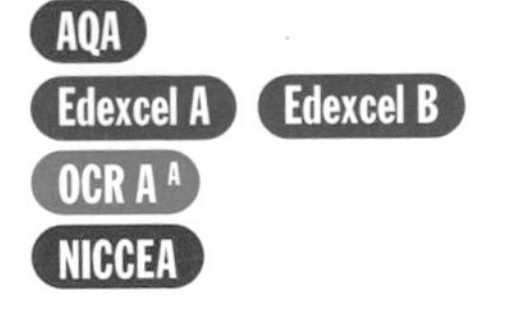

Ethanoic acid is a weak acid (see 5) with a molecular formula $C_2H_4O_2$.

It has a structural formula of

```
    H      O
    |    //
H - C - C
    |    \
    H     O
           \
            H
```

Ethanoic acid used to be called acetic acid. Vinegar is a dilute solution of ethanoic acid.

It is called ethanoic acid because, like ethane, it contains two carbon atoms.

Ethanoic acid is prepared in industry by passing ethanol and air over a heated catalyst.

Wine containing ethanol goes sour when left in contact with air. This is due to bacterial oxidation.

$$C_2H_5OH + O_2 \rightarrow CH_3COOH + H_2O$$

Properties of ethanoic acid

Ethanoic acid is a weak acid. It has similar reactions to other acids.

1. Indicators

Ethanoic acid turns litmus paper red. Solutions of ethanoic acid have a pH value of about 4.

2. Metals

Ethanoic acid reacts with reactive metals, e.g. magnesium to produce a salt plus **hydrogen**.

$Mg + 2CH_3COOH \rightarrow (CH_3COO)_2Mg + H_2$

magnesium + ethanoic acid $\rightarrow$ magnesium ethanoate + hydrogen

3. Metal oxides (bases)

Ethanoic acid reacts with a base to form a **salt** and **water** only.

$CuO + 2CH_3COOH \rightarrow (CH_3COO)_2Cu + H_2O$

copper(II) oxide + ethanoic acid $\rightarrow$ copper(II) ethanoate + water

4. Metal carbonates

Ethanoic acid reacts with a carbonate to produce a salt plus water and **carbon dioxide**.

$Na_2CO_3 + 2CH_3COOH \rightarrow 2CH_3COONa + H_2O + CO_2$

Sodium carbonate + ethanoic acid $\rightarrow$ sodium ethanoate + water + carbon dioxide

Methanoic acid, HCOOH, is a weak acid.

1. Draw the structural formula of methanoic acid.

Questions 2–8. Reagents are added to methanoic acid. Fill in the gaps in the table.

Test	Result of test	Name of substances formed
Litmus	2	
Add magnesium	3	4
Add copper(II) oxide	5	6
Add sodium carbonate	7	8

1. H–C(=O)–O–H

2. Turns red; 3. Bubbles of colourless gas; 4. Hydrogen and magnesium methanoate;
5. Blue solution; 6. Copper(II) methanoate and water; 7. Bubbles of colourless gas;
8. Sodium methanoate, carbon dioxide and water.

Esters

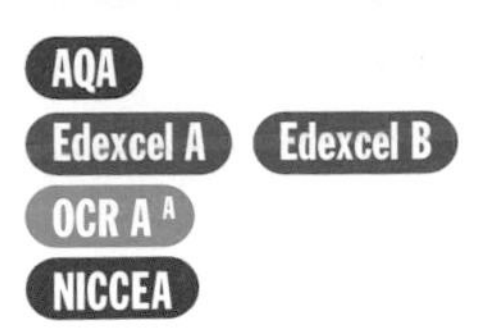

AQA
Edexcel A Edexcel B
OCR A A
NICCEA

KEY POINT **Esters are compounds formed by the reaction of an organic acid with an alcohol.**

Ethyl ethanoate is an ester produced by reaction of an acid (**ethanoic acid**) and an alcohol (**ethanol**).

$$CH_3COOH + C_2H_5OH \rightleftharpoons CH_3COOC_2H_5 + H_2O$$

ethanoic acid + ethanol ⇌ ethyl ethanoate + water

You will remember that acid + alkali → salt + water Now remember acid + alcohol ⇌ ester + water

A mixture of ethanol and ethanoic acid are heated under reflux (**Fig. 7.1** below) with a little concentrated sulphuric acid.

Heating under reflux prevents the reactants being lost by evaporation. If the mixture boils the vapour is condensed again and drops back into the flask.

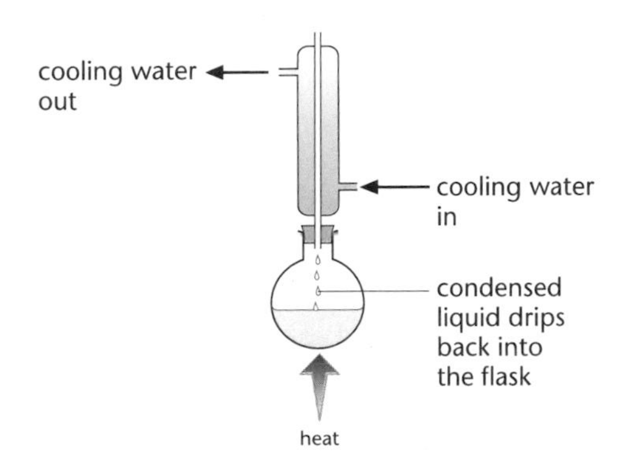

Fig. 7.1

Concentrated sulphuric acid removes the water and pushes the reaction to the right.

The structural formula of ethyl ethanoate is

```
    H     O
    |   //    H  H
H - C - C     |  |
    |    \O - C - C - H
    H         |  |
              H  H
```

Esters are **sweet-smelling liquids**. They are used as **flavouring agents**.

Ethyl ethanoate smells like pears and is used for flavouring sweets.

Natural fats and oils are naturally occurring esters derived from long chain acids e.g. $CH_3(CH_2)_{16}COOH$ and propane-1,2,3-triol (glycerol) $CH_2OHCHOHCH_2OH$.

The structural formula of a natural ester is

$$CH_2OOC(CH_2)_{16}CH_3$$
|
$$CHOOC(CH_2)_{16}CH_3$$
|
$$CH_2OOC(CH_2)_{16}CH_3$$

Fats and oils are used in the manufacture of soaps and margarine (see p. 83).

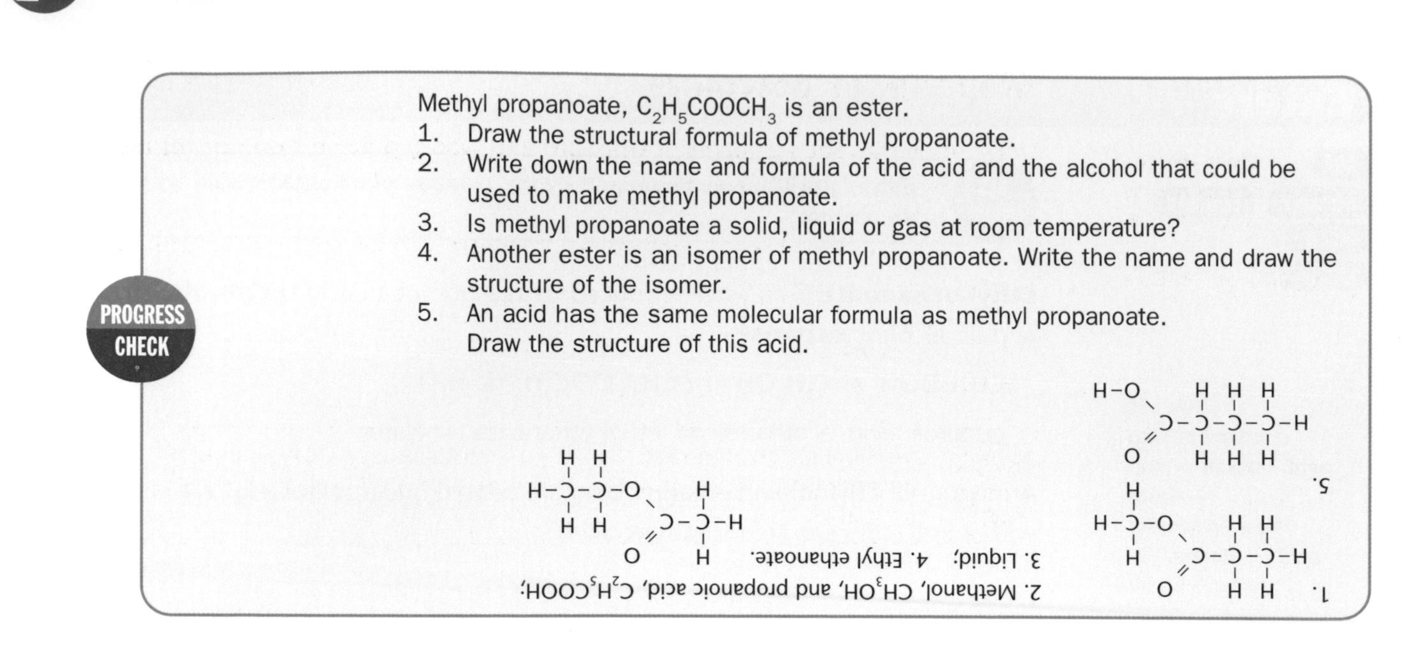

PROGRESS CHECK

Methyl propanoate, $C_2H_5COOCH_3$ is an ester.

1. Draw the structural formula of methyl propanoate.
2. Write down the name and formula of the acid and the alcohol that could be used to make methyl propanoate.
3. Is methyl propanoate a solid, liquid or gas at room temperature?
4. Another ester is an isomer of methyl propanoate. Write the name and draw the structure of the isomer.
5. An acid has the same molecular formula as methyl propanoate. Draw the structure of this acid.

2. Methanol, CH_3OH, and propanoic acid, C_2H_5COOH; 3. Liquid; 4. Ethyl ethanoate.

Further polymers

AQA
WJEC

Thermoplastic and thermosetting polymers

Polymers can be classified as **thermoplastic** or **thermosetting**.

KEY POINT

Thermoplastic polymers e.g. poly(ethene) melt easily when heated. Thermosetting polymers e.g. Bakelite do not melt when heated. On stronger heating they decompose.

There is a difference in structure between thermoplastic and thermosetting polymers.

A simple representation of the two types of polymer are shown in **Fig. 7.2** below. In a thermoplastic polymer the chains are not linked. On melting, the chains are able to move freely over each other. In a thermosetting polymer there are strong links between the polymer chains. The rigid structure is not easily broken down.

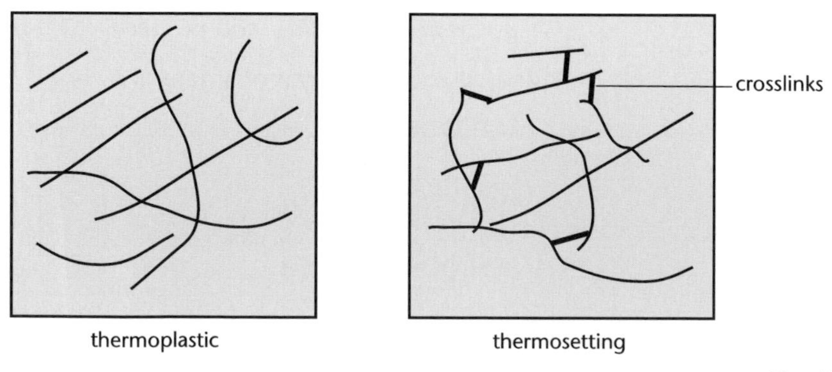

Fig. 7.2

Natural rubber consists of chains of polymer molecules. It is naturally soft and sticky. It can be hardened by a process of **vulcanisation** where sulphur atoms link the chains by cross-linking. The vulcanised rubber is now hard.

Disposal of polymers

Carbon dioxide produced increases the greenhouse effect.

Required for NICCEA

Unlike materials such as paper, cardboard and wood, polymers do not rot away when tipped in landfill sites. They are said to be **non-biodegradable**.

Recycling polymers is not economic as the costs of collection and sorting are greater than the costs of making new polymer.

Polymers can be **incinerated**. **Carbon dioxide** and **water vapour** are produced.

Some polymers can produce toxic gases: hydrogen cyanide from polymers containing nitrogen, hydrogen chloride from polymers containing chlorine.

1. A polymer melts when heated without decomposing. Is this polymer thermoplastic or thermosetting?
2. Which type of polymer, thermosetting or thermoplastic, is most suitable for
   (a) electric light fittings which must withstand heat;
   (b) plastic sheet which may be moulded to make trays;
   (c) heat-resistant saucepan handles;
   (d) plastic soldiers?

Three polymers are

poly(ethene) **poly(acrylonitrile)** **poly(chloroethene)**

3. Which polymer could produce hydrogen chloride on combustion?
4. Which polymer could produce hydrogen cyanide?
5. Why is it more difficult to recycle thermosetting polymers than thermoplastic polymers?

1. Thermoplastic; 2(a). Thermosetting; (b) Thermoplastic; (c) Thermosetting; (d) Thermoplastic; 3. Poly(chloroethene); 4. Poly(acrylonitrile); 5. They cannot be melted and remoulded without decomposition.

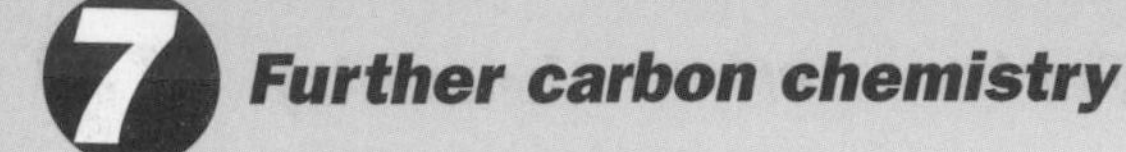

Sample GCSE questions

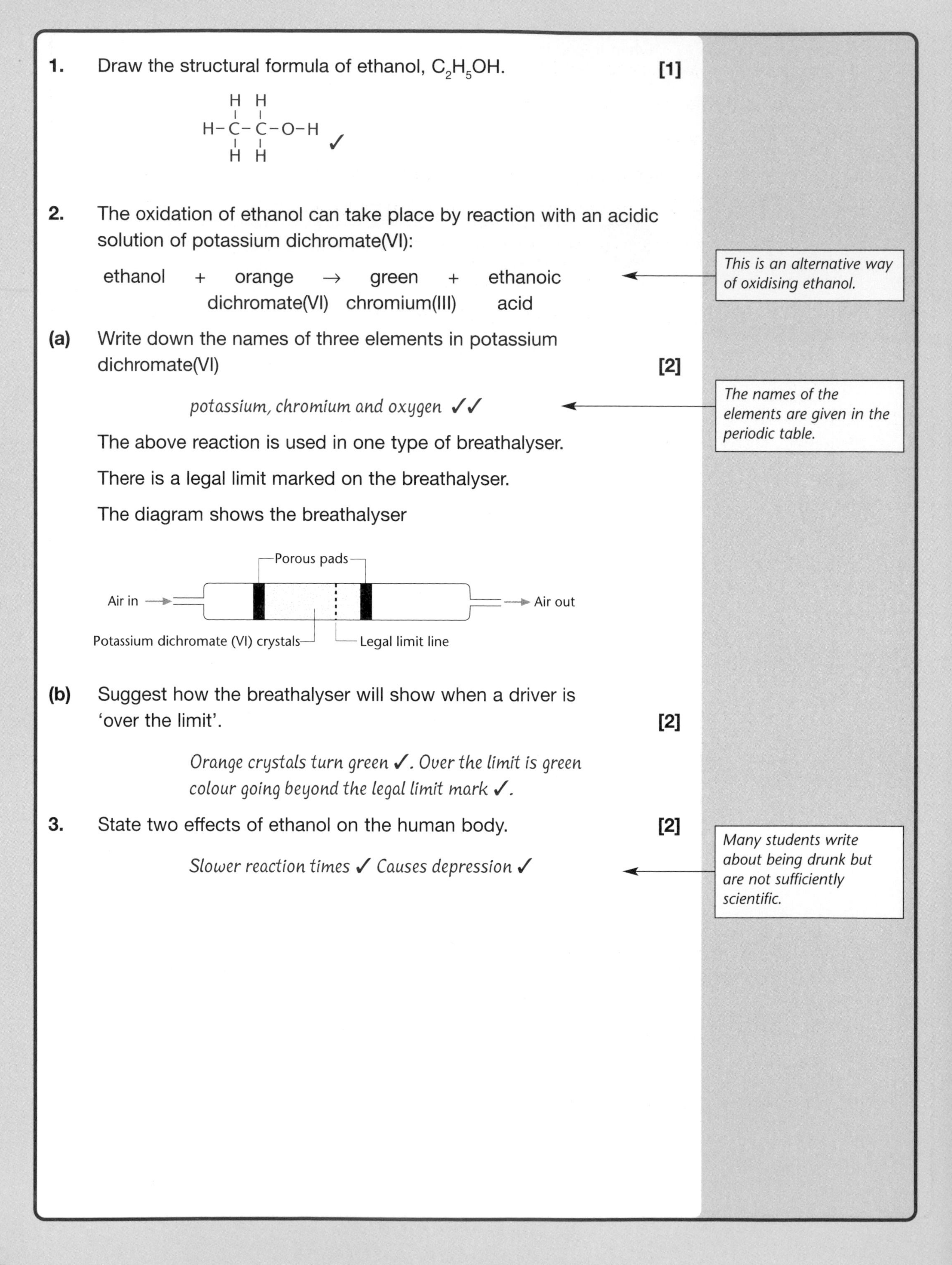

1. Draw the structural formula of ethanol, C_2H_5OH. **[1]**

```
    H   H
    |   |
H – C – C – O – H   ✓
    |   |
    H   H
```

2. The oxidation of ethanol can take place by reaction with an acidic solution of potassium dichromate(VI):

ethanol + orange dichromate(VI) → green chromium(III) + ethanoic acid

This is an alternative way of oxidising ethanol.

(a) Write down the names of three elements in potassium dichromate(VI) **[2]**

potassium, chromium and oxygen ✓✓

The names of the elements are given in the periodic table.

The above reaction is used in one type of breathalyser.

There is a legal limit marked on the breathalyser.

The diagram shows the breathalyser

(b) Suggest how the breathalyser will show when a driver is 'over the limit'. **[2]**

Orange crystals turn green ✓. Over the limit is green colour going beyond the legal limit mark ✓.

3. State two effects of ethanol on the human body. **[2]**

Slower reaction times ✓ Causes depression ✓

Many students write about being drunk but are not sufficiently scientific.

Exam practice question

1. Ethanol, C_2H_5OH, can be produced from ethene, C_2H_4, or from sugar.

Some ethanol is oxidised to produce ethanoic acid.

(a) Sugar is a carbohydrate.

(i) Write down the names of the three elements present in sugar. **[3]**

(ii) What is the name of the process used to turn sugar into ethanol? **[1]**

(iii) What must be present for the change from sugar to ethanol to take place? **[1]**

(b) Outline the processes used to produce ethene from crude oil. **[4]**

(c) Write down one use for ethene apart from making ethanol. **[1]**

(d) Write word and symbol equations for the reaction producing ethanol from ethene. **[3]**

(e) Low-grade wine containing ethanol can be turned into vinegar in the equipment shown in the diagram shown below.

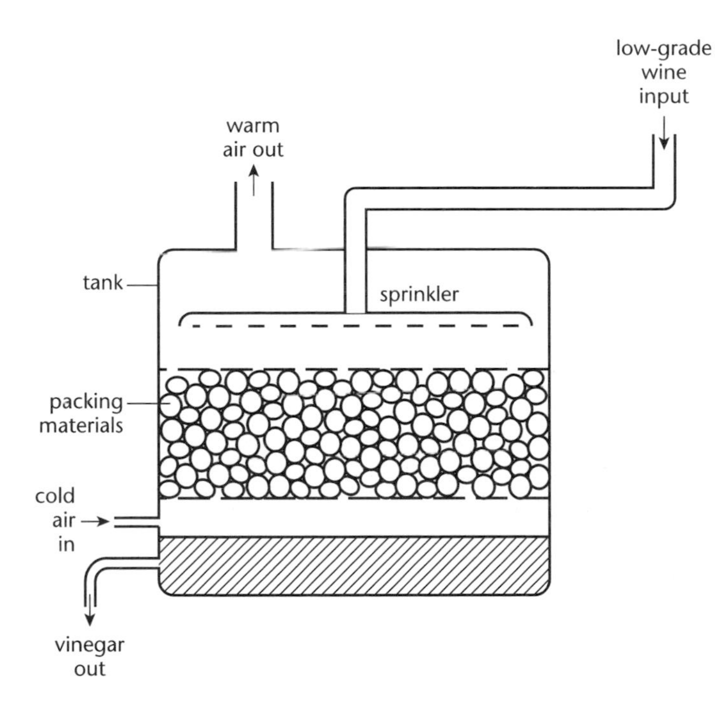

The wine is sprayed downward and air passes upwards over a packaging material. Bacteria, growing on the packing material, convert the ethanol in the wine into ethanoic acid.

The equation for the oxidation of ethanol into ethanoic acid is:

$C_2H_5OH(aq) + O_2(g) \rightarrow CH_3COOH(aq) + H_2O(l)$

(i) Why is this reaction an oxidation reaction? **[1]**

(ii) What evidence is there that the reaction is exothermic? **[1]**

Chapter

8 Further quantitative chemistry

The following topics are covered in this section:

- ***The mole***
- ***Calculating formulae from percentages***
- ***Volume changes in chemical reactions***
- ***Concentration of solutions***

LEARNING SUMMARY

After studying this section you should be able to:

- ***recall that the mole is the amount of substance containing Avogadro's number of particles***
- ***explain that one mole of any substance contains the same number of particles***
- ***work out the number of moles present in a given mass of substance***
- ***work out the mass in g of moles of different substances***
- ***work out formulae from percentages and percentages from formulae***
- ***work out volume changes during reactions involving gases***
- ***work out concentrations of solutions in mol/dm³***
- ***describe the process of titration in acid–alkali reactions.***

Chemical calculations are an important part of GCSE Chemistry Specifications. Many of these rely upon an understanding of the mole. Neutralisation reactions involving acids and alkalis are often carried out as titrations and from the results of these either concentrations or reacting volumes of solutions can be found.

The mole

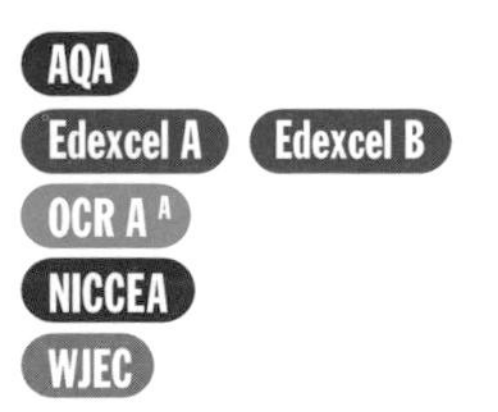

KEY POINT

The relative atomic mass of an atom is the number of times an atom is heavier than one twelfth of a carbon-12 atom. (See 2.4)

As an alternative to trying to consider masses of individual atoms, it is possible to compare a large numbers of atoms.

Atoms are so very small that individual atoms cannot be weighed.

The relative atomic mass of magnesium is twice the relative atomic mass of carbon.

1 atom of magnesium weighs twice as much as 1 atom of carbon-12. 2 atoms of magnesium weigh twice as much as 2 atoms of carbon-12. 100 atoms of magnesium weigh twice as much as 100 atoms of carbon-12.

The mass of magnesium atoms will always be twice the mass of the carbon-12 atoms, provided equal numbers of atoms are compared.

We are used to collective terms to describe a number of objects, e.g. a dozen eggs, a gross of test tubes, etc. In chemistry, the term **mole** (abbreviation mol) is used in the same way.

We refer to a mole of magnesium atoms, a mole of carbon dioxide molecules or a mole of electrons.

A mole is that amount of matter which contains 6×10^{23} particles (600 000 000 000 000 000 000 000). This number is called Avogadro's constant (L).

Avogadro's number is so large that if the whole population of the world wished to count up to this number between them and they worked at counting without breaks it would take six million years to finish.

A mole of atoms of any element has a mass equal to the relative atomic mass (but with units of grams).

One mole of magnesium atoms has a mass of 24g and one mole of carbon atoms has a mass of 12g. Notice that the mass of magnesium is still twice the mass of carbon, because equal numbers of particles are considered.

KEY POINT

The mole may be defined as the amount of substance which contains as many elementary units as there are atoms in 12g of carbon.

A mole of carbon atoms is a small handful. As there are 6×10^{23} atoms in the pile, it emphasises just how small atoms are.

These elementary units can be considered as:

atoms e.g. Mg, C, He

molecules e.g. CH_4, H_2O

ions e.g. Na^+, Cl^-

specified formula units e.g. H_2SO_4

The term '1 mole of chlorine' can be ambiguous. It could mean 1 mole of chlorine atoms (6×10^{23} atoms) or 1 mole of chlorine molecules, Cl_2 (12×10^{23} atoms).

Calculating the mass of 1 mole

Calculate the mass of:

(a) 1 mole of oxygen atoms, O

(b) 1 mole of oxygen molecules, O_2

(c) 1 mole of methane molecules, CH_4

(d) 1 mole of sulphuric acid, H_2SO_4

If you look up relative atomic masses on the periodic table (page 54) make sure you use relative atomic masses rather than atomic numbers.

Relative atomic masses $A_r(H) = 1$, $A_r(C) = 12$, $A_r(O) = 16$, $A_r(S) = 32$

(a) 16g

This is the relative atomic mass of oxygen in g.

(b) 16 + 6 = 32g

(c) $12 + (4 \times 1) = 16$g

(d) $(2 \times 1) + 32 + (4 \times 16) = 98$g

Calculating the number of moles in given masses

$$\textbf{number of moles} = \frac{\textbf{mass in g}}{\textbf{mass of 1 mole in g}}$$

Calculate the number of moles represented by:

(a) 8g of oxygen molecules

(b) 8g of methane molecules

(c) 9.8g of sulphuric acid

(d) 22g of carbon dioxide, CO_2

(a) Number of moles = $\frac{8}{32}$ = 0.25

(b) Number of moles = $\frac{8}{16}$ = 0.5

(c) Number of moles = $\frac{9.8}{98}$ = 0.1

(d) Mass of 1 mole of CO_2 = 12 + (2 × 16) = 44g

Number of moles = $\frac{22}{44}$ = 0.5

This formula can be rearranged

mass in g = number of moles × mass of 1 mole in g

Calculate the mass of:

(a) 0.1 moles of methane

(b) 2 moles of sulphuric acid

(c) 0.5 moles of sulphur dioxide, SO_2

(a) Mass = 0.1 × 16 = 1.6g

(b) Mass = 2 × 98 = 196g

(c) Mass of 1 mole of sulphur dioxide = 32 + (2 × 16) = 64g

Mass = 0.5 × 64 = 32g

Volume of 1 mole of gas

There is no simple relationship that predicts the volume occupied by 1 mole of molecules in a solid or liquid.

For gases

This information will be given on examination papers if it is needed.

1 mole of any gas occupies 24 000 cm³ (or 24 dm³) at room temperature and atmospheric pressure.

Calculate the volume of 0.1 moles of carbon dioxide at room temperature and atmospheric pressure.

Answer = 24 000 × 0.1 = 2400 cm³

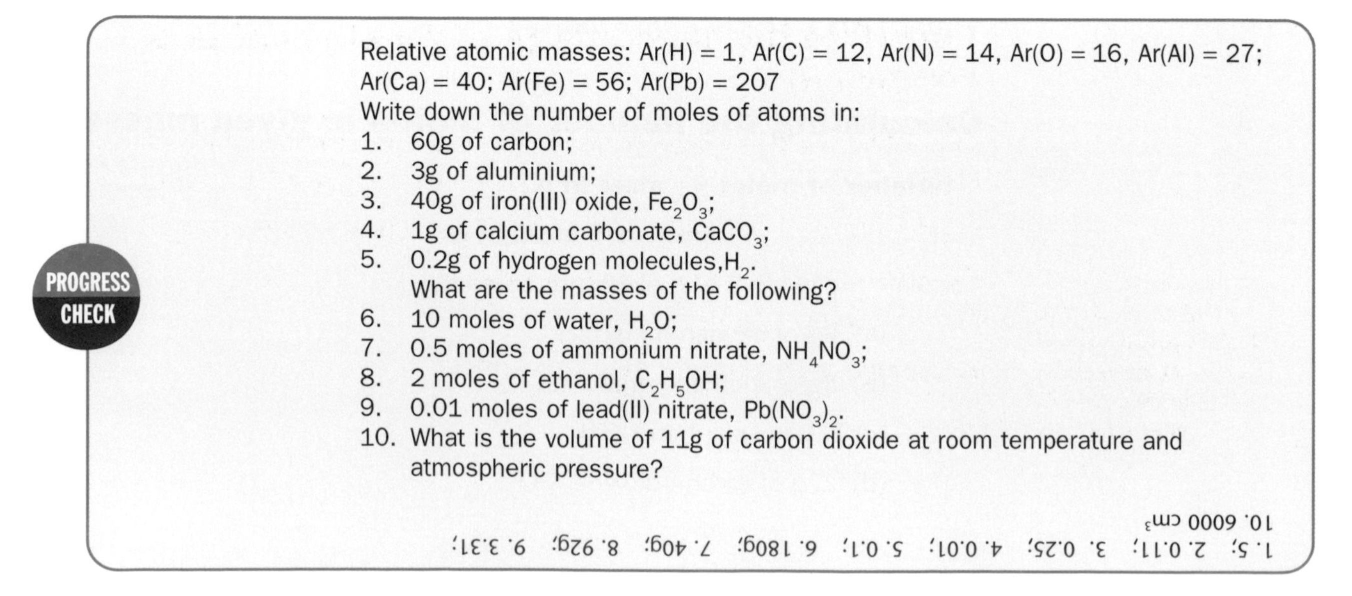

PROGRESS CHECK

Relative atomic masses: Ar(H) = 1, Ar(C) = 12, Ar(N) = 14, Ar(O) = 16, Ar(Al) = 27; Ar(Ca) = 40; Ar(Fe) = 56; Ar(Pb) = 207

Write down the number of moles of atoms in:

1. 60g of carbon;
2. 3g of aluminium;
3. 40g of iron(III) oxide, Fe_2O_3;
4. 1g of calcium carbonate, $CaCO_3$;
5. 0.2g of hydrogen molecules, H_2.

What are the masses of the following?

6. 10 moles of water, H_2O;
7. 0.5 moles of ammonium nitrate, NH_4NO_3;
8. 2 moles of ethanol, C_2H_5OH;
9. 0.01 moles of lead(II) nitrate, $Pb(NO_3)_2$.
10. What is the volume of 11g of carbon dioxide at room temperature and atmospheric pressure?

1. 5; 2. 0.11; 3. 0.25; 4. 0.01; 5. 0.1; 6. 180g; 7. 40g; 8. 92g; 9. 3.31;
10. 6000 cm³

Calculating formulae from percentages

AQA
Edexcel A Edexcel B
OCR A A
NICCEA
WJEC

Chemical formulae were worked out using relative atomic masses on page 42 (2.4)

Chemical formulae can be worked out using moles.

e.g. 4.14g of lead combines with 0.64g of oxygen. (A_r(Pb) = 207; A_r(O) =16)

Number of moles of lead = 4.14 ÷ 207 = 0.02

Number of moles of oxygen = 0.64 ÷ 16 = 0.04

A common error here is to write Pb_2O.

There are twice as many particles of oxygen as lead, so the simplest formula is PbO_2.

You can use percentages of different elements to work out the formula.

Calculate the formula of a compound containing iron, sulphur and oxygen. It contains 28% iron and 24% sulphur.

Remember that the percentages of all the elements add up to 100

Relative atomic masses : A_r(Fe) = 56; A_r(S) = 32; A_r(O) = 16

Percentage of oxygen in the compound = 100 – (28 + 24) = 48%

Number of moles of iron = 28 ÷ 56 = 0.5

Number of moles of sulphur = 24 ÷ 32 = 0.75

Number of moles of oxygen = 48 ÷ 16 = 3

0.25 divides into 0.5, 0.75 and 3.

Divide each by 0.25

The simplest formula is $Fe_2S_3O_{12}$

The simplest formula is called the **empirical formula**.

KEY POINT **The molecular formula is either the empirical formula or some multiple of it.**

The empirical formula of a compound is CH_2. The mass of 1 mole of the compound is 56g.

The molecular formula of the compound is 4 × the empirical formula i.e. C_4H_8.

This has a mass of 1 mole of 4 × 14 = 56g.

Calculate the percentage of an element in a compound

Calculate the percentage of potassium in potassium hydrogencarbonate, $KHCO_3$.

Relative atomic masses A_r(H) = 1, A_r(C) = 12, A_r(O) = 16 , A_r(K) = 39.

These calculations are often used to calculate percentage of nitrogen, phosphorus or potassium in a fertiliser.

Mass of 1 mole of potassium hydrogencarbonate = 39 + 1 + 12 + (3×16) = 100g.

Percentage of potassium = $\frac{39}{100} \times 100 = 39\%$.

PROGRESS CHECK

A hydrocarbon X contains 75% carbon.

1. Calculate the percentage of hydrogen in the hydrocarbon.
2. Calculate the empirical (simplest) formula of the hydrocarbon.

Relative atomic masses $A_r(H) = 1$, $A_r(C) = 12$.

3. The mass of 1 mole of the hydrocarbon is 16g. What is the molecular formula?
4. Suggest a hydrocarbon that could be X.
5. What is the percentage of nitrogen in ammonium nitrate, NH_4NO_3? Relative atomic masses $A_r(H) = 1$, $A_r(N) = 14$, $A_r(O) = 16$.

1. 25%; 2. CH_4; 3. CH_4; 4. Methane; 5. 35%.

Volume changes in chemical reactions

AQA
Edexcel A Edexcel B
OCR A A
NICCEA
WJEC

On page 41 (2.4) it was shown how masses of chemicals reacting, and masses of products formed could be worked out using a balanced symbol equation.

e.g. The equation for the burning of carbon in excess oxygen is

$$C(s) + O_2(g) \rightarrow CO_2(g)$$

The equation shows us that 1 mole of carbon (12g) reacts with 1 mole of oxygen (32g) to produce 1 mole of carbon dioxide.

Remember the sum of the mass of the reactants is the same as the sum of the mass of the products. The number of moles on each side is not the same.

We know that

1 mole of molecules of any gas occupies 24 000 cm^3 (or 24 dm^3) at room temperature and atmospheric pressure.

12g of carbon react with 24 dm^3 of oxygen (at room temperature and atmospheric pressure) to form 24 dm^3 of carbon dioxide.

Another example:

hydrogen + oxygen → water

$$2H_2(g) + O_2(g) \rightarrow 2H_2O(l)$$

4g of hydrogen reacts with 32g of oxygen to form 36g of water.

In terms of moles

2 moles of hydrogen molecules reacts with 1 mole of oxygen molecules to form 2 moles of water.

At room temperature and atmospheric pressure

48 dm^3 of hydrogen reacts with 24 dm^3 of oxygen. As steam has condensed to liquid water, the volume of the products is negligible.

There is a huge decrease in volume during the reaction.

The sum of the volumes of the reactants is not necessarily equal to the sum of the volumes of the products.

PROGRESS CHECK

1. In which of these reactions is the volume of the reactants equal to the volume of the products?
   A. $CH_4(g) + 2O_2(g) \rightarrow CO_2(g) + 2H_2O(l)$
   B. $H_2(g) + Cl_2(g) \rightarrow 2HCl(g)$
   C. $2CH_4(g) + 3O_2(g) \rightarrow 2CO(g) + 4H_2O(l)$
   D. $N_2(g) + 3H_2(g) \rightarrow 2NH_3(g)$

The equation for the complete combustion of ethane is

$2C_2H_6(g) + 7O_2(g) \rightarrow 4CO_2(g) + 6H_2O(l)$

2. What volume of oxygen is needed to react with 30 cm^3 of ethane, C_2H_6?
3. What volume of carbon dioxide would be produced by the complete combustion of 20 cm^3 of ethane.

1. B; 2. 105 cm^3; 3. 40 cm^3.

Concentration of solutions

AQA · Edexcel A · Edexcel B · OCR A[A] · NICCEA · WJEC

Measuring the concentration of a solution

The concentration of a solution can be measured in units of g/dm^3.

For example, if 10g of salt are dissolved in 100g of water, the concentration of the salt solution is $100g/dm^3$.

g/dm^3 and g/litre are the same. Sometimes they are written as $g\ dm^{-3}$ or $g\ l^{-3}$.

Measuring concentration in this way gives no comparison of the number of particles in a given volume of solution.

When 1 mole of solute is dissolved in water and the solution made up to a volume of 1 dm^3 the solution has a concentration of 1 mole/dm^3. This is usually written as 1 mol/dm^3 and sometimes called a molar (or M) solution.

Equal volumes of two solutions each 1 mol/dm^3 will contain the same number of particles.

e.g. 8g of sodium hydroxide solution is dissolved in water and made up to a volume of 100 cm^3 with water.

(a) What is the concentration of the solution in mol/dm^3?

(b) How many moles of sodium hydroxide are present in 25cm^3 of this solution?

Relative atomic masses: $A_r(H) = 1, A_r(O) = 16, A_r(Na) = 23$.

(a) 1 mole of NaOH = 23 + 16 + 1 = 40g

Number of moles of NaOH = 8 ÷ 40 = 0.2 moles

Concentration of the solution is 0.2 moles/100cm^3 or 2 mol/dm^3

(b) 25cm^3 of solution is one fortieth of one dm^3, so contains 2 ÷ 40 moles i.e. 0.05 moles.

Finding the concentration of another solution by titration

This is a neutralisation reaction and the ionic equation is $H^+ + OH^- \rightarrow H_2O$

Sodium hydroxide solution and dilute sulphuric acid react to form sodium sulphate.

$$2NaOH(aq) + H_2SO_4(aq) \rightarrow Na_2SO_4(aq) + 2H_2O$$

The equation shows that 2 moles of sodium hydroxide react with 1 mole of sulphuric acid to form one mole of sodium sulphate and two moles of water.

If 25 cm^3 of sodium hydroxide (0.1 mol/dm^3) are reacted with sulphuric acid (0.1 mol/dm^3), 12.5 cm^3 of sulphuric acid would be needed.

Fig. 8.1 below summarises the steps used for a **titration** to find the volumes of acid and alkali exactly reacting to produce a neutral solution.

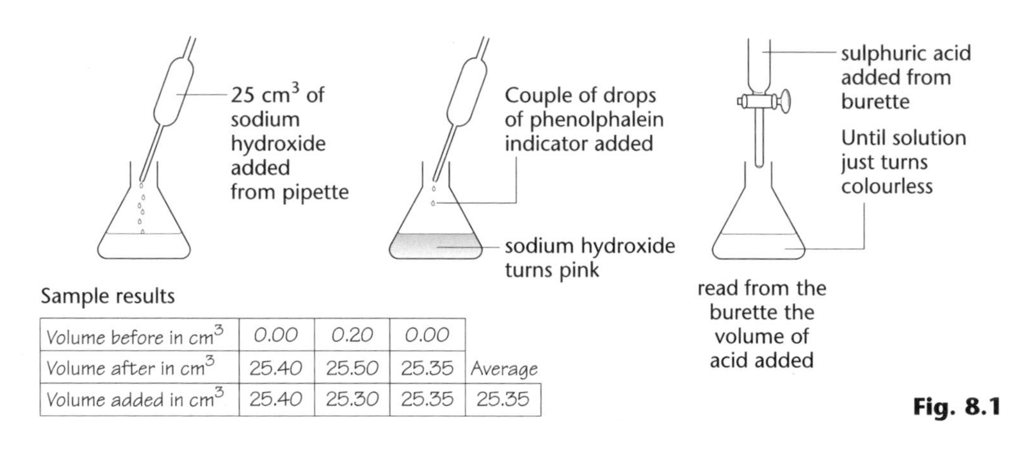

Sample results

Volume before in cm^3	0.00	0.20	0.00	
Volume after in cm^3	25.40	25.50	25.35	Average
Volume added in cm^3	25.40	25.30	25.35	25.35

Fig. 8.1

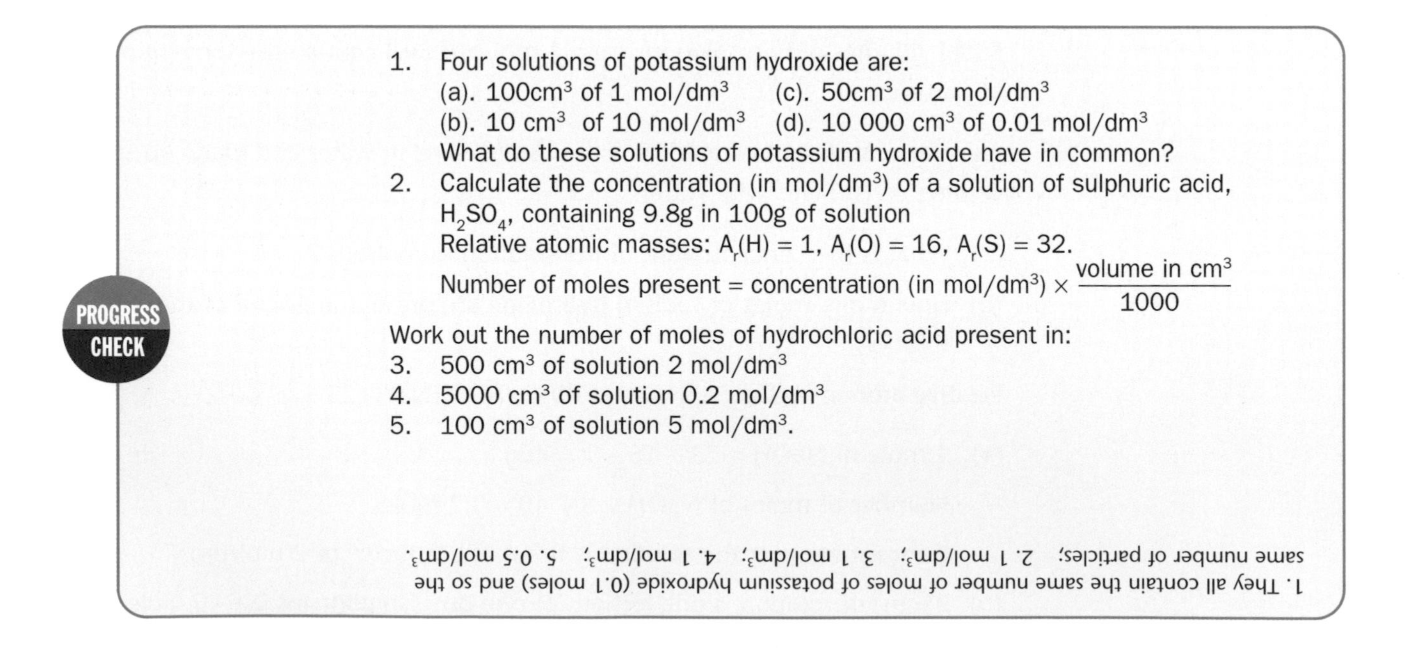

PROGRESS CHECK

1. Four solutions of potassium hydroxide are:
   (a). 100cm^3 of 1 mol/dm^3 (c). 50cm^3 of 2 mol/dm^3
   (b). 10 cm^3 of 10 mol/dm^3 (d). 10 000 cm^3 of 0.01 mol/dm^3
   What do these solutions of potassium hydroxide have in common?
2. Calculate the concentration (in mol/dm^3) of a solution of sulphuric acid, H_2SO_4, containing 9.8g in 100g of solution
   Relative atomic masses: $A_r(H) = 1$, $A_r(O) = 16$, $A_r(S) = 32$.

$$\text{Number of moles present} = \text{concentration (in mol/dm}^3) \times \frac{\text{volume in cm}^3}{1000}$$

Work out the number of moles of hydrochloric acid present in:

3. 500 cm^3 of solution 2 mol/dm^3
4. 5000 cm^3 of solution 0.2 mol/dm^3
5. 100 cm^3 of solution 5 mol/dm^3.

1. They all contain the same number of moles of potassium hydroxide (0.1 moles) and so the same number of particles; 2. 1 mol/dm^3; 3. 1 mol/dm^3; 4. 1 mol/dm^3; 5. 0.5 mol/dm^3.

Sample GCSE question

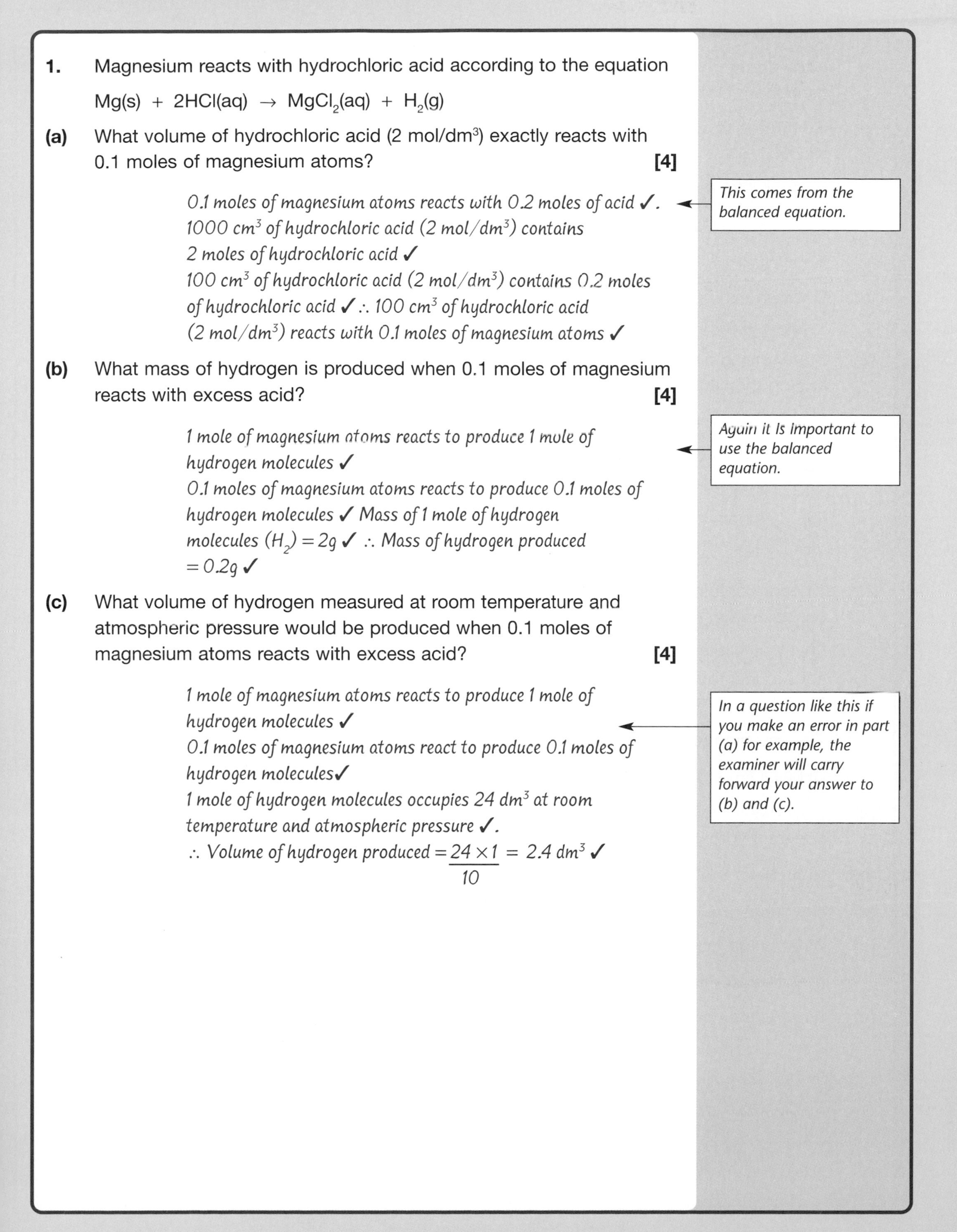

1. Magnesium reacts with hydrochloric acid according to the equation

$Mg(s) + 2HCl(aq) \rightarrow MgCl_2(aq) + H_2(g)$

(a) What volume of hydrochloric acid (2 mol/dm³) exactly reacts with 0.1 moles of magnesium atoms? **[4]**

0.1 moles of magnesium atoms reacts with 0.2 moles of acid ✓. 1000 cm³ of hydrochloric acid (2 mol/dm³) contains 2 moles of hydrochloric acid ✓ 100 cm³ of hydrochloric acid (2 mol/dm³) contains 0.2 moles of hydrochloric acid ✓ ∴ 100 cm³ of hydrochloric acid (2 mol/dm³) reacts with 0.1 moles of magnesium atoms ✓

This comes from the balanced equation.

(b) What mass of hydrogen is produced when 0.1 moles of magnesium reacts with excess acid? **[4]**

1 mole of magnesium atoms reacts to produce 1 mole of hydrogen molecules ✓ 0.1 moles of magnesium atoms reacts to produce 0.1 moles of hydrogen molecules ✓ Mass of 1 mole of hydrogen molecules (H_2) = 2g ✓ ∴ Mass of hydrogen produced = 0.2g ✓

Again it is important to use the balanced equation.

(c) What volume of hydrogen measured at room temperature and atmospheric pressure would be produced when 0.1 moles of magnesium atoms reacts with excess acid? **[4]**

1 mole of magnesium atoms reacts to produce 1 mole of hydrogen molecules ✓ 0.1 moles of magnesium atoms react to produce 0.1 moles of hydrogen molecules ✓ 1 mole of hydrogen molecules occupies 24 dm³ at room temperature and atmospheric pressure ✓.

∴ Volume of hydrogen produced $= \frac{24 \times 1}{10} = 2.4\ dm^3$ ✓

In a question like this if you make an error in part (a) for example, the examiner will carry forward your answer to (b) and (c).

Exam practice questions

1. Andy carried out an experiment to check the concentration of some sulphuric acid being used in an electroplating process.

He titrated 25.0 cm^3 of the acid with sodium hydroxide solution containing 1.2 mol/dm^3 NaOH.

He found 35.0 cm^3 of the alkali was required for neutralisation.

The equation for the reaction is:

$2NaOH(aq) + H_2SO_4(aq) \rightarrow Na_2SO_4(aq) + 2H_2O(l)$

(a) What should he use to measure out **(i)** 25.0 cm^3 of acid; **(ii)** the alkali in small measured portions? **[2]**

(b) State, giving a named example, what he added in order to know when to stop adding the alkali. **[2]**

(c) **(i)** How many moles of sodium hydroxide (NaOH) were added in the titration? **[1]**

(ii) With how many moles of sulphuric acid, H_2SO_4, did 35.0 cm^3 of sodium hydroxide react? **[1]**

(iii) What is the concentration of the sulphuric acid in mol/dm^3? **[1]**

2. Vinegar contains ethanoic acid. An experiment was carried out to find the concentration of ethanoic acid in a sample of white vinegar.

50cm^3 of vinegar was put into a flask. Sodium hydroxide solution (0.1 mol/dm^3) was added until the indicator changed colour. The volume of sodium hydroxide added was 20cm^3.

The equation for the reaction is:

$CH_3COOH + NaOH \rightarrow CH_3COONa + H_2O$

(a) How many moles of sodium hydroxide are present in 20cm^3 of 0.1 mol/dm^3 sodium hydroxide solution? **[2]**

(b) How many moles of ethanoic acid are present in 50cm^3 of vinegar? **[1]**

(c) How many moles of ethanoic acid are present in 1000cm^3 of vinegar? **[1]**

(d) What is the mass of 1 mole of ethanoic acid?

(Relative atomic masses Ar (H) = 1, Ar (C) = 12, Ar (O) = 16) **[1]**

(e) What is the concentration of ethanoic acid, in g/dm^3 in vinegar? **[2]**

Chapter 9 Electrochemistry and electrolysis

The following topics are covered in this section:

- Electrical conductivity
- Electrolysis of molten zinc chloride
- Electrolysis of aqueous solutions
- Uses of electrolysis
- Quantitative electrolysis

LEARNING SUMMARY

After studying this section you should be able to:

- recall that metals and graphite conduct electricity when solid without change because they contain free moving electrons
- recall that electrolytes such as sodium chloride conduct electricity and are decomposed when molten or in solution
- explain what is happening during the electrolysis of molten zinc chloride
- explain the products of electrolysis of some aqueous solutions of electrolytes and be able to predict others
- work out masses of products of electrolysis.

Electrolysis is an important process. It was developed by Michael Faraday in the early nineteenth century when suitable supplies of electricity became available. He discovered a number of new elements, e.g. the alkali metals and established the basic laws of electrolysis.

Electrical conductivity

AQA
OCR A [A] OCR A [B]
NICCEA

Fig. 9.1 below shows apparatus that can be used to test if a substance conducts electricity. If electricity passes through X the bulb lights.

The only common examples of solids conducting electricity are metals and graphite (a form of carbon).
Electricity passes through metals and graphite because of freely moving (delocalised) electrons.
Electricity passes through metals and graphite with no chemical change taking place.

Liquid metals also conduct electricity and again no decomposition takes place.

Solid acids, alkalis and salts (called **electrolytes**) do not conduct electricity. Although they contain ions, the ions are held in fixed positions and are not able to move.

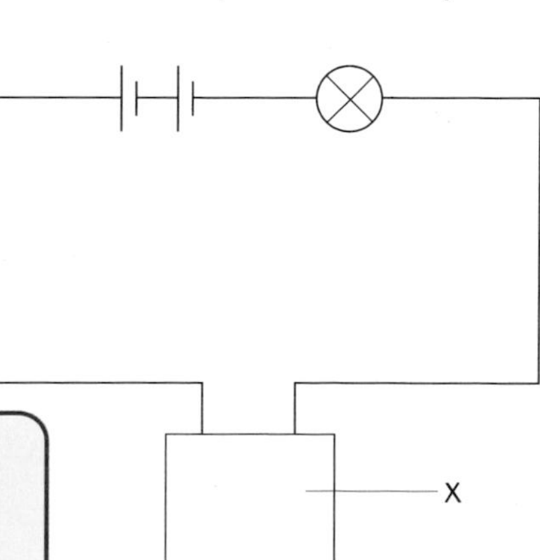

Fig. 9.1

Melting an electrolyte or dissolving it in water breaks up the structure and the ions are free to move.
Electrolysis is the splitting up of an electrolyte when molten or in solution.

Electrolytes contain **ionic bonds**.

KEY POINT Ions move (or migrate) towards the electrode of opposite charge.

This is shown in **Fig. 9.2.**

When the ions reach the electrode they may be **discharged**. This involves either a transfer of electrons to or from the electrode.

Compounds that are liquid or in solution but do not conduct electricity contain **covalent bonds**.

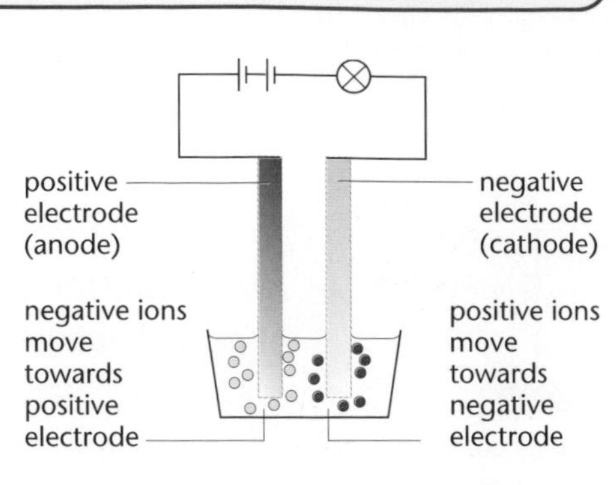

Fig. 9.2

PROGRESS CHECK

copper **ethanol** **graphite** **mercury**
molten sodium chloride **poly(ethene)** **sodium**
sodium chloride solution **solid sodium chloride**

1. Which of the substances in the list will conduct electricity?
2. Which of the substances in the list will conduct electricity without decomposition?

1. Copper, graphite, mercury, molten sodium chloride, sodium , sodium chloride solution;
2. Copper, graphite, mercury, sodium

Electrolysis of molten zinc chloride

AQA OCR A[A] OCR A[B] NICCEA

The apparatus in **Fig. 9.3** could be used for the electrolysis of zinc chloride, $ZnCl_2$:

- the bulb does not light while the zinc chloride is solid
- as soon as the zinc chloride melts, the bulb lights
- zinc starts to form around the negative electrode
- chlorine gas is produced at the positive electrode
- the bulb goes out when heat is removed and the melt solidifies.

The overall reaction is

$ZnCl_2 \rightarrow Zn + Cl_2$

The carbon rods are called **electrodes**. The electrode attached to the positive terminal of the battery is the positive electrode (sometimes called the **anode**).

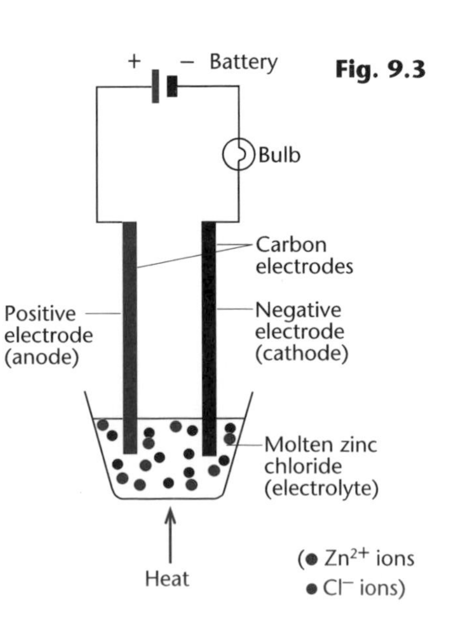

Fig. 9.3

The electrode attached to the negative terminal of the battery is the negative electrode (sometimes called the **cathode**).

The positive electrode has a shortage of electrons and the negative electrode has a surplus of electrons. Electrons are constantly flowing through the wire from the positive electrode to the negative electrode.

Migration of ions

When the zinc chloride is melted, **positive** zinc ions, Zn^{2+}, move towards the **negative** electrode.

Ions move towards the electrode of opposite charge.

The **negative** chloride ions, Cl^-, move towards the positive electrode.

Discharging of ions

When the ions arrive at the electrodes they are **discharged**.

Remember the cathode has an excess of electrons. This process involves a gain of electrons and so is an example of reduction.

At the negative electrode (cathode) electrons are transferred from the cathode to the ion and the ion is changed to a metal atom.

$Zn^{2+} + 2e \rightarrow Zn$

Remember the anode has a shortage of electrons and this change is providing more electrons. This process involves a loss of electrons and so is an example of oxidation.

At the positive electrode (anode) electrons are transferred from the ion to the anode and the ion is changed to a chlorine atom. Two chlorine atoms combine to form a chlorine molecule.

$Cl^- \rightarrow Cl + e^-$ $2Cl \rightarrow Cl_2$

PROGRESS CHECK

1. In an experiment to electrolyse molten zinc chloride the electrolysis was carried out for a long time. On cooling the bulb continued to glow brightly. Explain why this might happen.

A similar experiment was carried out with lithium iodide, LiI, used in place of zinc chloride.

2. Which ions are present in lithium iodide?
3. Which ion moves towards the cathode?
4. Which ion moves towards the anode?
5. What is produced at the cathode?
6. What is produced at the anode?
7. Write an ionic equation for the change at the cathode.
8. Write an ionic equation for the change at the anode.

1. The zinc produced links the anode and cathode and electricity passes through it without going through the melt; 2. Li^+ and I^-; 3. Li^+; 4. I^-; 5. Lithium; 6. Iodine;
7. $Li^+ + e^- \rightarrow Li$; $I^- \rightarrow I + e^-$ and $I + I \rightarrow I_2$ or $2I^- \rightarrow I_2 + 2e^-$

Electrolysis of aqueous solutions

AQA OCR A[A] OCR A[B] NICCEA

In pure water about 1 in every 600 000 000 water molecules is split into hydrogen and hydroxide ions.

$$H_2O(l) \rightleftharpoons H^+(aq) + OH^-(aq)$$

This slight ionisation of water explains why pure water has a low electrical conductivity.

Electrolysis of sodium chloride solution

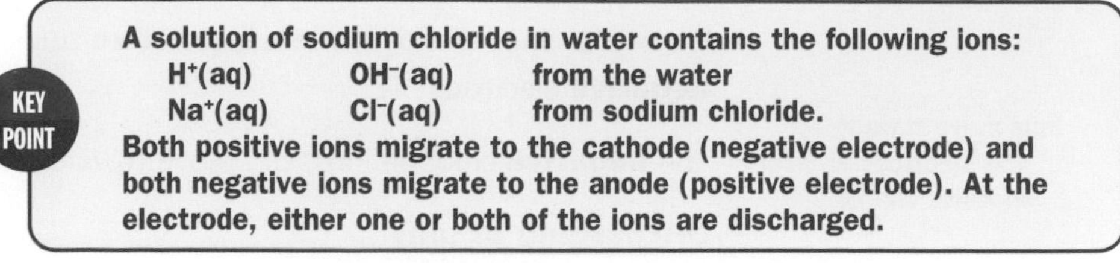

KEY POINT

A solution of sodium chloride in water contains the following ions:

$H^+(aq)$ $OH^-(aq)$ from the water

$Na^+(aq)$ $Cl^-(aq)$ from sodium chloride.

Both positive ions migrate to the cathode (negative electrode) and both negative ions migrate to the anode (positive electrode). At the electrode, either one or both of the ions are discharged.

Table 9.1 below gives the results of some electrolysis experiments.

Solution	Electrodes	Ion discharged at positive electrode	Product at at positive electrode	Ion discharged at negative electrode	Product at negative electrode
Dilute sulphuric acid	carbon	$OH^-(aq)$	oxygen	$H^+(aq)$	hydrogen
Dilute sodium hydroxide	carbon	$OH^-(aq)$	oxygen	$H^+(aq)$	hydrogen
Copper(II) sulphate	carbon	$OH^-(aq)$	oxygen	Cu^{2+} (aq)	copper
Copper(II) sulphate	copper	none	none	Cu^{2+} (aq)	copper
Copper(II) chloride	carbon	$Cl^-(aq)$	chlorine	Cu^{2+} (aq)	copper
Very dilute sodium chloride	carbon	$OH^-(aq)$	oxygen	$H^+(aq)$	hydrogen
Concentrated sodium chloride	carbon	$Cl^-(aq)$	chlorine	$H^+(aq)$	hydrogen
Concentrated sodium chloride	mercury cathode	$Cl^-(aq)$	chlorine	$Na^+(aq)$	sodium amalgam

The apparatus in **Fig. 9.4** can be used for the electrolysis of aqueous solutions. It enables the gases produced at the electrodes to be collected.

- **Metals**, if produced, are produced at the **negative** electrode.
- **Hydrogen** is produced at the **negative** electrode only.
- **Non-metals**, apart from hydrogen, are produced at the **positive** electrode.
- **Reactive metals** are not formed at the positive electrode during the electrolysis of aqueous solutions. An exception occurs during the electrolysis of sodium chloride using a mercury cathode.

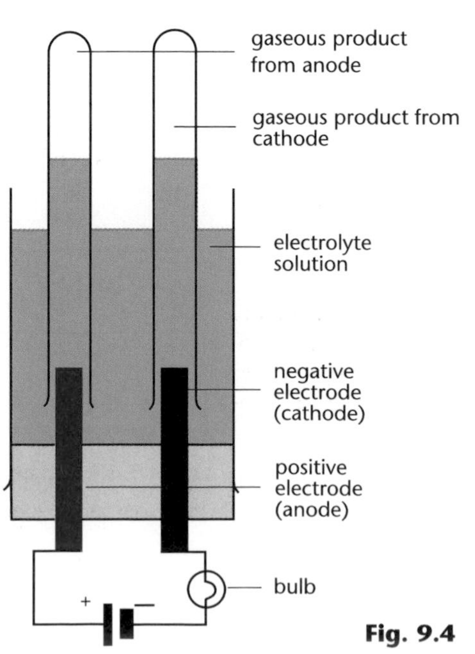

Fig. 9.4

- The products can depend upon the concentration of the electrolyte in the solution. For example, electrolysis of concentrated sodium chloride solution produces chlorine at the positive electrode but the electrolysis of dilute sodium chloride can produce oxygen.
- Providing the concentrations of the negative ions in solution are approximately the same, the order of discharge is

$OH^-(aq)$
$I^-(aq)$
$Br^-(aq)$
$Cl^-(aq)$
$NO_3^-(aq)$
$SO_4^{2-}(aq)$

↓ ease of discharge decreases

PROGRESS CHECK

1. Which ions are present in dilute sodium hydroxide solution?
2. Which of these ions is not discharged when sodium hydroxide solution is electrolysed?
3. When copper(II) sulphate is electrolysed with copper electrodes no gas is produced at the positive electrode. The electrode does, however, reduce in size. Suggest how electrons are produced at this electrode.
4. Which ions are present in a concentrated solution of potassium iodide, KI, in water?
5. What are the products of the electrolysis of a concentrated solution of potassium iodide?
6. Suggest how it might be different if a very dilute solution of potassium iodide was used.

1. Na^+, H^+, OH^-; 2. Na^+; 3. $Cu \rightarrow Cu^{2+} + 2e^-$; 4. K^+, H^+, I^-, OH^-; 5. Iodine (at positive electrode) and hydrogen (at the negative electrode); 6. Oxygen may be collected instead of iodine.

Uses of electrolysis

Electrolysis is used widely in industry.

KEY POINT

Some uses include:
- **extraction of reactive metals such as sodium and aluminium (2.2)**
- **purification of copper (2.2)**
- **electroplating of metals**
- **electrolysis of brine.**

Electroplating of metals

This can be done for decorative purposes or to prevent corrosion.

Electrolysis can be used to put a thin coating of a metal on the surface of another metal.

If a piece of copper is to be plated with a layer of nickel, the copper must first be thoroughly cleaned and dried. **Copper** is made the **cathode** and a piece of **nickel** the **anode**. They are both dipped into **nickel(II) sulphate solution** (the **electrolyte**).

Nickel is deposited on the copper cathode.

$$Ni^{2+} + 2e^- \rightarrow Ni$$

The nickel anode goes into solution as nickel ions.

$$Ni \rightarrow Ni^{2+} + 2e^-$$

Electrolysis of brine

Electrolysis of brine (sodium chloride solution) produces sodium hydroxide, hydrogen and chlorine.

Fig. 9.5 shows a diaphragm cell into which concentrated brine is pumped.

In the cell, reactions occur at the electrodes.

At the titanium anode

$$2Cl^-(aq) \rightarrow Cl_2(g) + 2e^-$$

At the steel cathode

$$2H^+(aq) + 2e^- \rightarrow H_2(g)$$

If sodium hydroxide and chlorine react together sodium chlorate(I) is produced. This is used as household bleach.

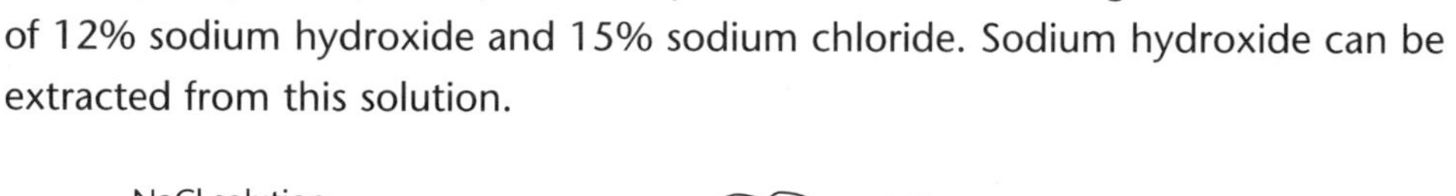

The diaphragm keeps the products apart. The solution leaving the cell consist of 12% sodium hydroxide and 15% sodium chloride. Sodium hydroxide can be extracted from this solution.

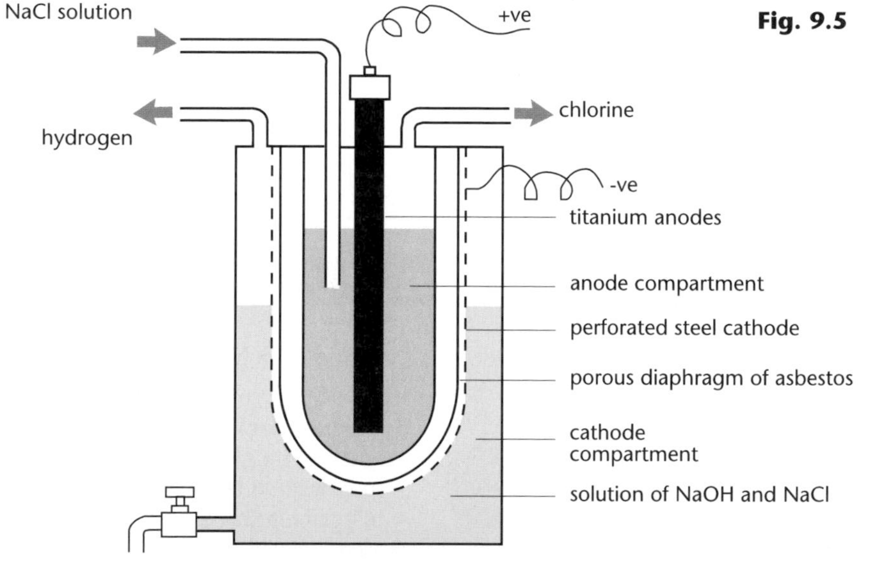

Fig. 9.5

PROGRESS CHECK

A brooch made of a copper alloy is going to be silver plated.

1. What is the name of the process that can be used to do this?
2. Why is this better than just dipping it in molten silver?
3. Is the brooch made the anode or cathode in the process?
4. Suggest a metal for the other electrode.
5. Suggest a suitable electrolyte.
6. Write an equation for each electrode process.

1. Electroplating; 2. Gives a thinner layer or a more even layer; 3. Cathode; 4. Silver; 5. A soluble silver salt e.g. silver nitrate; 6. $Ag^+ + e^- \rightarrow Ag$; $Ag \rightarrow Ag^+ + e^-$.

Quantitative electrolysis

Calculating the quantity of electricity

KEY POINT **The quantity of electricity used in an electrolysis experiment can be calculated using the formula Q = I × t.**

Remember to turn the time into seconds. If it is in minutes multiply by 60.

In this equation the quantity of electricity, Q, is measured in coulombs, the current passing, I, is measured in amperes and the time the electricity is passing, t, is measured in seconds.

A coulomb is a very small quantity of electricity.

Example: calculate the quantity of electricity passed when a current of 4A passes for 2 minutes.

Quantity of electricity $= 4 \times (2 \times 60) = 480$ coulombs

1 faraday is the charge carried by a mole of electrons.

A faraday is a larger unit for measuring the quantity of electricity.

One faraday = 96 000 coulombs.

The unit faraday is named after Michael Faraday who developed the theory of electrolysis at the beginning of the nineteenth Century.

In the example above, 480 coulombs is equivalent to $480 \div 96\,000$ faradays = 0.005 faradays

Calculating the mass deposited by a quantity of electricity

The mass of product formed at an electrode is dependent upon the quantity of electricity.

This is Faraday's First Law of electrolysis. The mass of a given product is directly proportional to the quantity of electricity.

If 0.005 faradays of electricity produces 0.54g of silver in an experiment, 0.01 faradays (twice the quantity) produces 1.08g of silver (twice the mass).

Table 9.2 below gives the quantity of electricity required to produce 1 mole of atoms of different elements.

Ion being discharged	No. of faradays to discharge 1 mole of atoms
Na^+	1
Ca^{2+}	2
Cu^{2+}	2
Al^{3+}	3
Cl^-	1
O^{2-}	2

If the ions being discharged at the positive or negative electrodes have a single charge (positive or negative), 1 faraday of electricity will discharge 1 mole of atoms.
If the ions have a double charge, 2 faradays will discharge 1 mole of atoms. Similarly, if the ions have a triple charge, 3 faradays will discharge 1 mole of atoms.

Example: How many faradays of electricity are needed to discharge 0.1 moles of iodine atoms from iodide, I^- ions?

If you want to find the charge on an ion, find out how many faradays of electricity are required to produce 1 mole of atoms during electrolysis.

Since iodide has a single negative charge:

1 faraday is required to discharge 1 mole of iodine atoms

∴ 0.1 faradays are required to discharge 0.1 moles of iodine atoms.

PROGRESS CHECK

1. Calculate the number of coulombs of electricity passed when a current of 3A passes for 1 hour.
2. Calculate the number of faradays of electricity passed when a current of 1.6A passes for 600 seconds.
3. Calculate the mass of calcium produced when 0.2 faradays of electricity are passed through molten calcium bromide, $CaBr_2$. (A_r(Ca)=40)
4. How many faradays are required to produce 1 mole of aluminium atoms during the extraction of aluminium from Al_2O_3?

1. 10 800 coulombs; 2. 0.01 faradays; 3. 4.0g; 4. 3.

Sample GCSE question

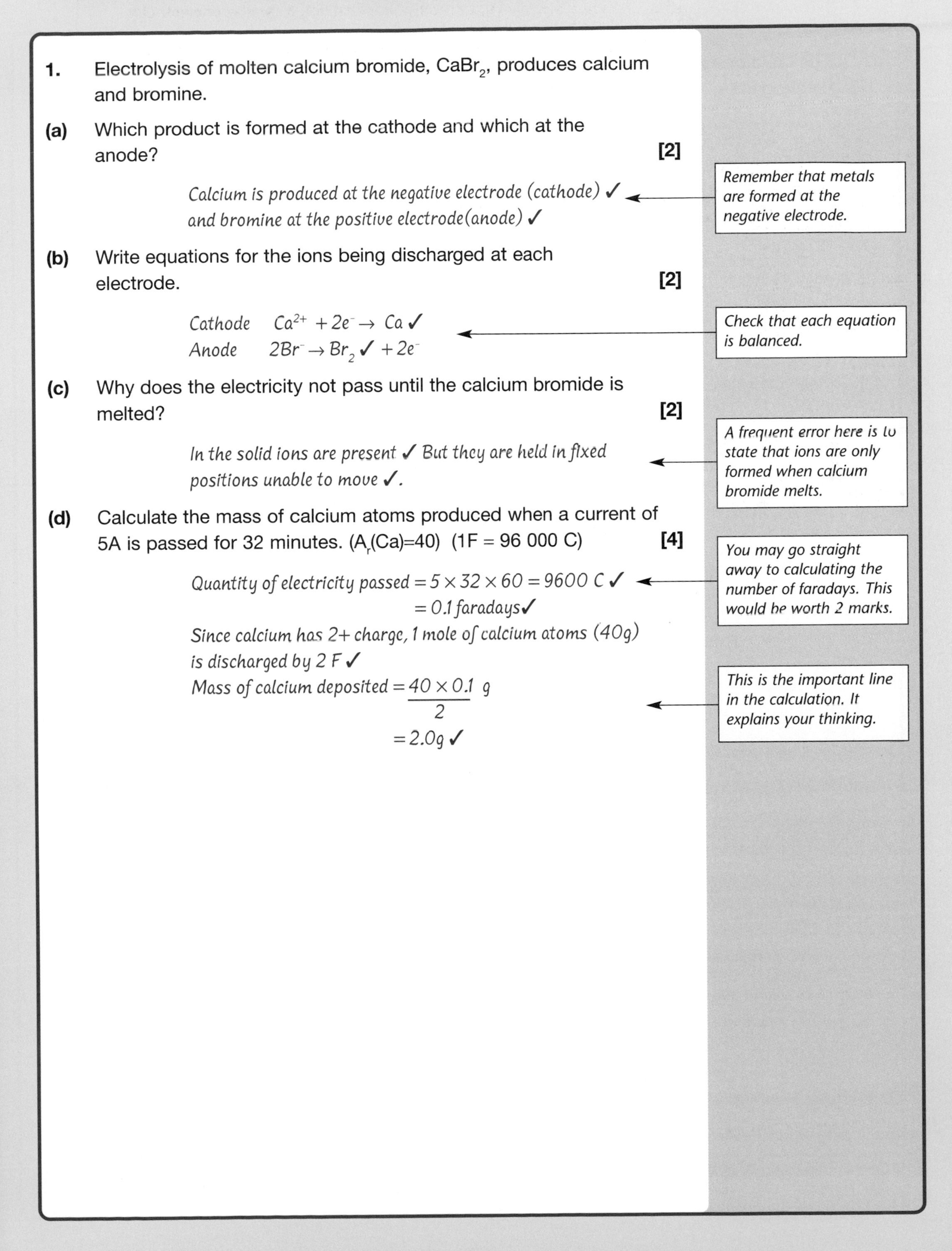

1. Electrolysis of molten calcium bromide, $CaBr_2$, produces calcium and bromine.

(a) Which product is formed at the cathode and which at the anode? **[2]**

Calcium is produced at the negative electrode (cathode) ✓ and bromine at the positive electrode(anode) ✓

Remember that metals are formed at the negative electrode.

(b) Write equations for the ions being discharged at each electrode. **[2]**

Cathode $Ca^{2+} + 2e^- \rightarrow Ca$ ✓
Anode $2Br^- \rightarrow Br_2$ ✓ $+ 2e^-$

Check that each equation is balanced.

(c) Why does the electricity not pass until the calcium bromide is melted? **[2]**

In the solid ions are present ✓ But they are held in fixed positions unable to move ✓.

A frequent error here is to state that ions are only formed when calcium bromide melts.

(d) Calculate the mass of calcium atoms produced when a current of 5A is passed for 32 minutes. (A_r(Ca)=40) (1F = 96 000 C) **[4]**

Quantity of electricity passed $= 5 \times 32 \times 60 = 9600$ C ✓
$= 0.1$ *faradays* ✓
Since calcium has 2+ charge, 1 mole of calcium atoms (40g) is discharged by 2 F ✓
Mass of calcium deposited $= \dfrac{40 \times 0.1}{2}$ g
$= 2.0$g ✓

You may go straight away to calculating the number of faradays. This would be worth 2 marks.

This is the important line in the calculation. It explains your thinking.

Exam practice question

1. The table gives the products of the electrolysis of some electrolytes with carbon electrodes.

Electrolyte	Ion discharged at positive electrode	Product at positive electrode	Ion discharged at negative electrode	Product at negative electrode
Molten sodium chloride	Cl^-	chlorine	Na^+	sodium
Sodium chloride solution (conc)				
Sodium sulphate solution				
Copper(II) sulphate solution				

(a) Finish the table. **[6]**

(b) How would the products be different if copper(II) sulphate solution was electrolysed with copper electrodes? **[3]**

Chapter 10

Collection of gases

The following topics are covered in this section:

- ***Methods of collecting gases***
- ***Tests for gases***

LEARNING SUMMARY

After studying this section you should be able to:

- ***recall methods used to collect gases and suggest a suitable method to collect a given gas***
- ***recall the tests for some common gases.***

KEY POINT

One of the huge advances in the eighteenth century was to be able to handle gases, usually by collection over water or over mercury. Today it is important that you can collect gases produced in chemical reactions and identify them with chemical tests.

Methods of collecting gases

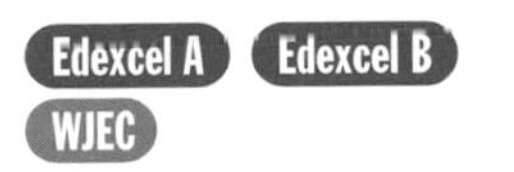

Edexcel A Edexcel B WJEC

The method used to collect a gas depends upon properties of the gas.

Gases insoluble or slightly soluble in water

Gases that are insoluble in water or not very soluble in water can be collected **over water** (see **Fig. 10.1**).

Lavoisier collected gases over mercury. This would work with gases that are soluble in water but the mercury is very toxic and too expensive to use.

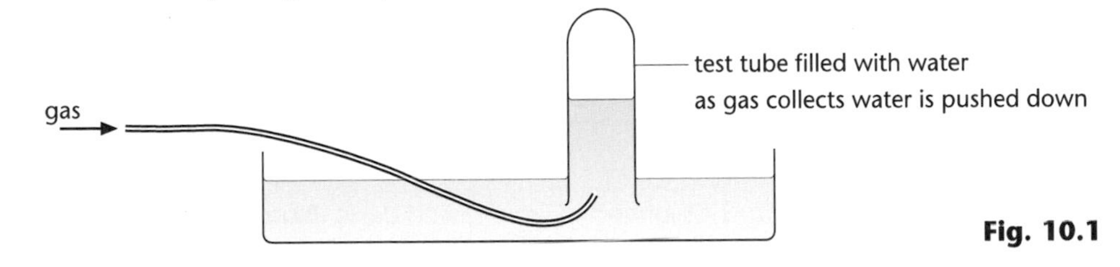

Fig. 10.1

The test tube is filled with water and as the gas is collected the water is pushed out of the tube. If the volume of the gas is needed, a measuring cylinder or burette can be used in place of the test tube.

Gases with high and low density

Gases with a **low density** (lighter than air) are collected by **upward delivery** (**Fig. 10.2** below). As the gas is collected air is pushed out of the tube.

Gases with a **high density** (heavier than air) are collected by **downward delivery** (**Fig. 10.3**). As the gas is collected air is pushed out of the tube.

If the gas collected is colourless it is difficult to tell when the tube is full.

All gases can be collected in a gas syringe (3.3).

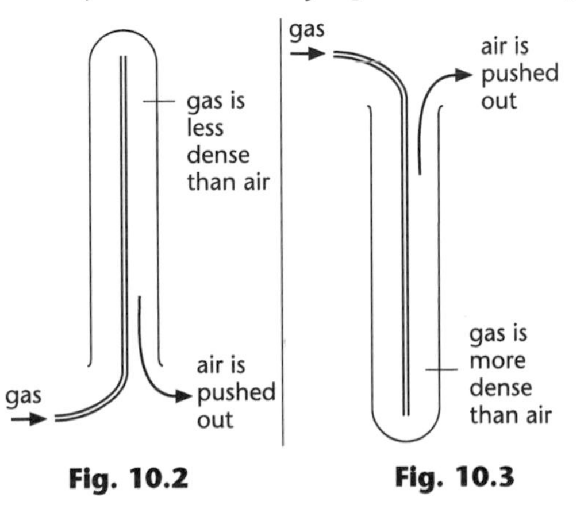

Fig. 10.2 **Fig. 10.3**

PROGRESS CHECK

The table gives some properties of gases labelled A, B, C and D

Gas	Solubility in water	Density in g/dm^3
A	Low	0.08
B	High	0.71
C	High	1.16
D	High	2.66

Density of air under the same conditions 1.21 g/dm^3

1. Which gas could be collected by downward delivery?
2. Which gas could be collected by upward delivery?
3. Which gas could be collected over water?
4. Which gas could not be collected over water or by upward or downward delivery?

1. D; 2. B; 3. A; 4. C.

Tests for gases

AQA
Edexcel A Edexcel B
OCR A [A] OCR A [B]
NICCEA
WJEC

Table 10.1 below gives the tests for common gases.

Gas	Test	Positive result
Oxygen	Put a **glowing** split into gas	Splint **relights**
Hydrogen	Put a **lighted** splint into gas	**Squeaky pop** and splint extinguished
Ammonia	Add damp **red litmus paper**	Turns **blue**
Carbon dioxide*	Bubble through **limewater**	Limewater turns **milky**
Hydrogen chloride	Open tube of **ammonia** gas and let gases mix	Dense **white fumes** formed.
Sulphur dioxide	Bubble through **orange potassium dichromate solution**	Solution turns **green**
Chlorine*	Add damp **blue litmus paper**	Turns **red** and then **bleaches**.

* Gases only required by WJEC.

PROGRESS CHECK

1. Which gas turns litmus paper blue?
2. Which gas turns limewater milky?
3. A gas turns damp blue litmus paper red and turns potassium dichromate solution green. Which gas is this?
4. A gas is mixed with ammonia gas and dense white fumes are formed. What is this gas?

1. Ammonia; 2. Carbon dioxide; 3. Sulphur dioxide; 4. Hydrogen chloride.

Sample GCSE question

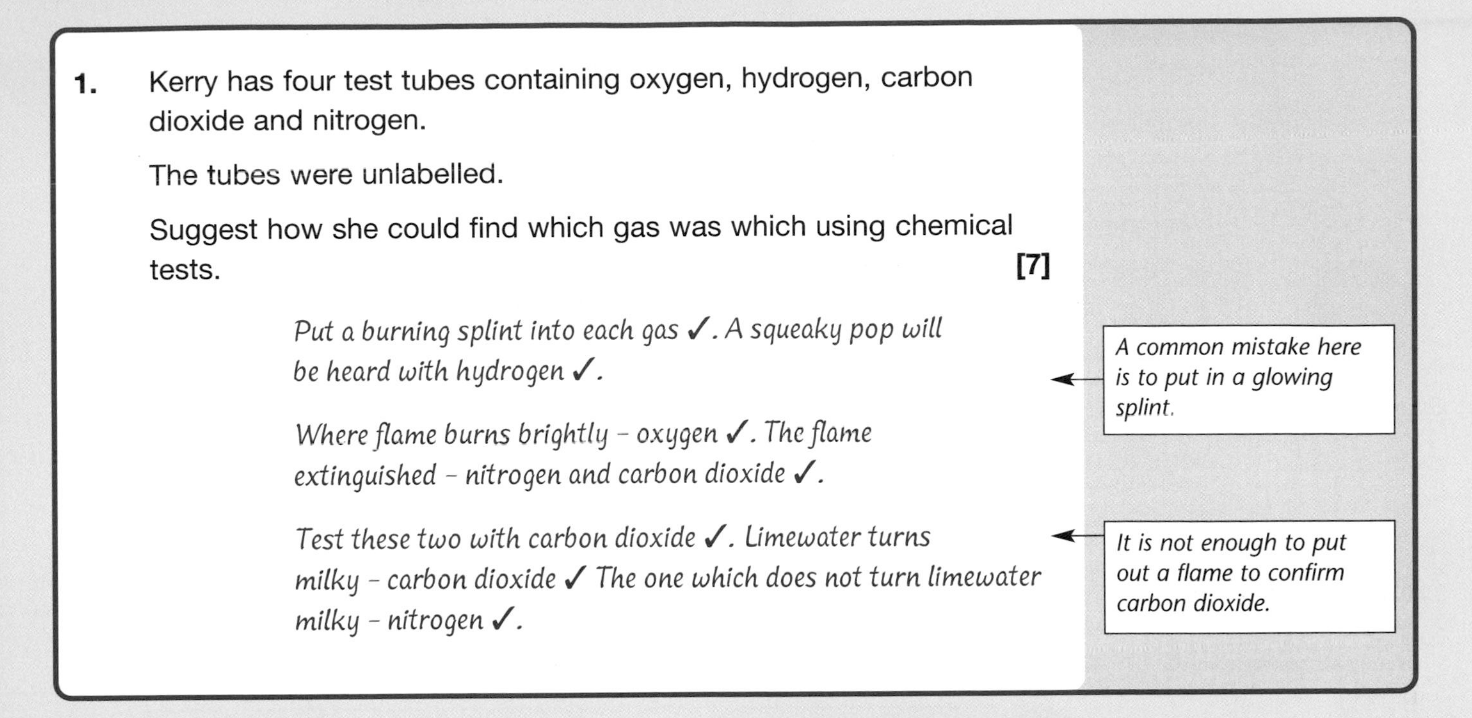

1. Kerry has four test tubes containing oxygen, hydrogen, carbon dioxide and nitrogen.

The tubes were unlabelled.

Suggest how she could find which gas was which using chemical tests. **[7]**

Put a burning splint into each gas ✓. A squeaky pop will be heard with hydrogen ✓.

Where flame burns brightly – oxygen ✓. The flame extinguished – nitrogen and carbon dioxide ✓.

Test these two with carbon dioxide ✓. Limewater turns milky – carbon dioxide ✓ The one which does not turn limewater milky – nitrogen ✓.

A common mistake here is to put in a glowing splint.

It is not enough to put out a flame to confirm carbon dioxide.

Exam practice question

1. When a mixture of solid ammonium chloride and sodium hydroxide is heated in a test tube ammonia gas is produced.

(a) Finish the symbol equation

$NH_4Cl + NaOH \rightarrow NH_3 +$ **[1]**

(b) Ammonia is collected by upward delivery.

(i) Suggest one reason why it is not collected over water. **[1]**

(ii) How could you show when a test tube collecting ammonia is full of ammonia? **[3]**

Food and drugs

The following topics are covered in this section:

- ***Carbohydrates, proteins and fats***
- ***Drugs***
- ***Vitamins and food additives***

LEARNING SUMMARY

After studying this section you should be able to:

- ***recall that carbohydrates, proteins and fats are needed in a balanced diet***
- ***describe the structure of carbohydrates, proteins and fats***
- ***recall that food contains vitamins and food additives***
- ***recall that certain drugs are able to relieve pain.***

A balanced diet contains carbohydrates, proteins and fats. There are also smaller quantities of vitamins, minerals, etc. Other chemicals such as food additives and raising agents may be added.

Carbohydrates, proteins and fats

OCR A B

KEY POINT

Carbohydrates are compounds containing carbon, hydrogen and oxygen fitting a general formula $C_x(H_2O)_y$.

Carbohydrates may be divided into **monosaccharides** (e.g. glucose and fructose, both $C_6H_{12}O_6$), **disaccharides** (e.g. sucrose and maltose, both $C_{12}H_{22}O_{11}$) and **polysaccharides** (e.g. starch and cellulose). Polysaccharides are polymers of monosaccharides.

Fig. 11.1 below shows two structures – a chain structure and a ring structure. The ring structure better explains the structure of glucose.

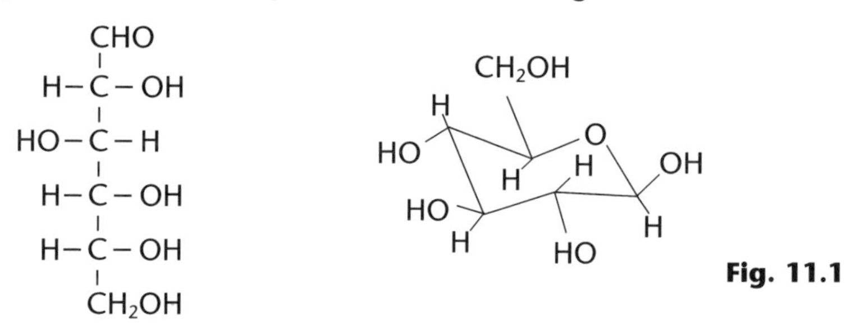

Fig. 11.1

Sucrose is the carbohydrate we commonly call sugar. A reaction where two molecules are joined and a small molecule lost is called a condensation reaction.

Sucrose is formed when a **glucose** and a **fructose** molecule join together with the loss of a water molecule.

Fig. 11.2 shows a simple representation of a sucrose molecule.

Starch and cellulose are **condensation polymers (Fig. 11.3).**

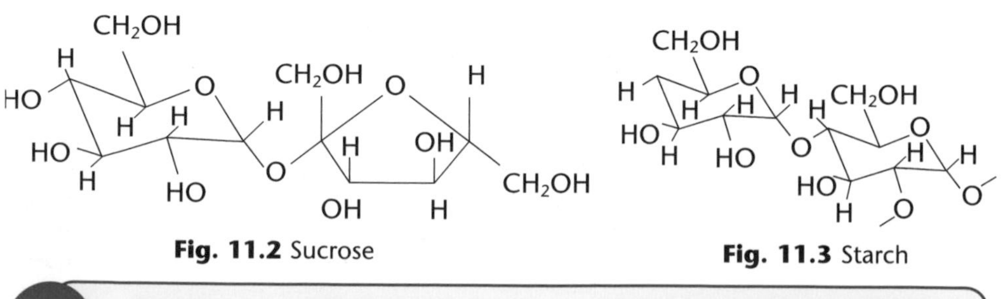

Fig. 11.2 Sucrose

Fig. 11.3 Starch

Proteins are condensation polymers made by linking together a large number of amino acid molecules.

Fig. 11.4 shows how two amino acid molecules are linked together by a **peptide link** to form a **dipeptide**.

A protein is made when many amino acid molecules are joined together by peptide links.

```
    H  O     H
    |  ||    |
R - C - C - N - C - R'
    |       |   |
   NH2      H  COOH
```

Fig. 11.4

A single protein can contain as many as 500 amino acid units combined together. A protein can be pictured as in **Fig. 11.5** below as coils with the loops of the coils held in position by weak crosslinks.

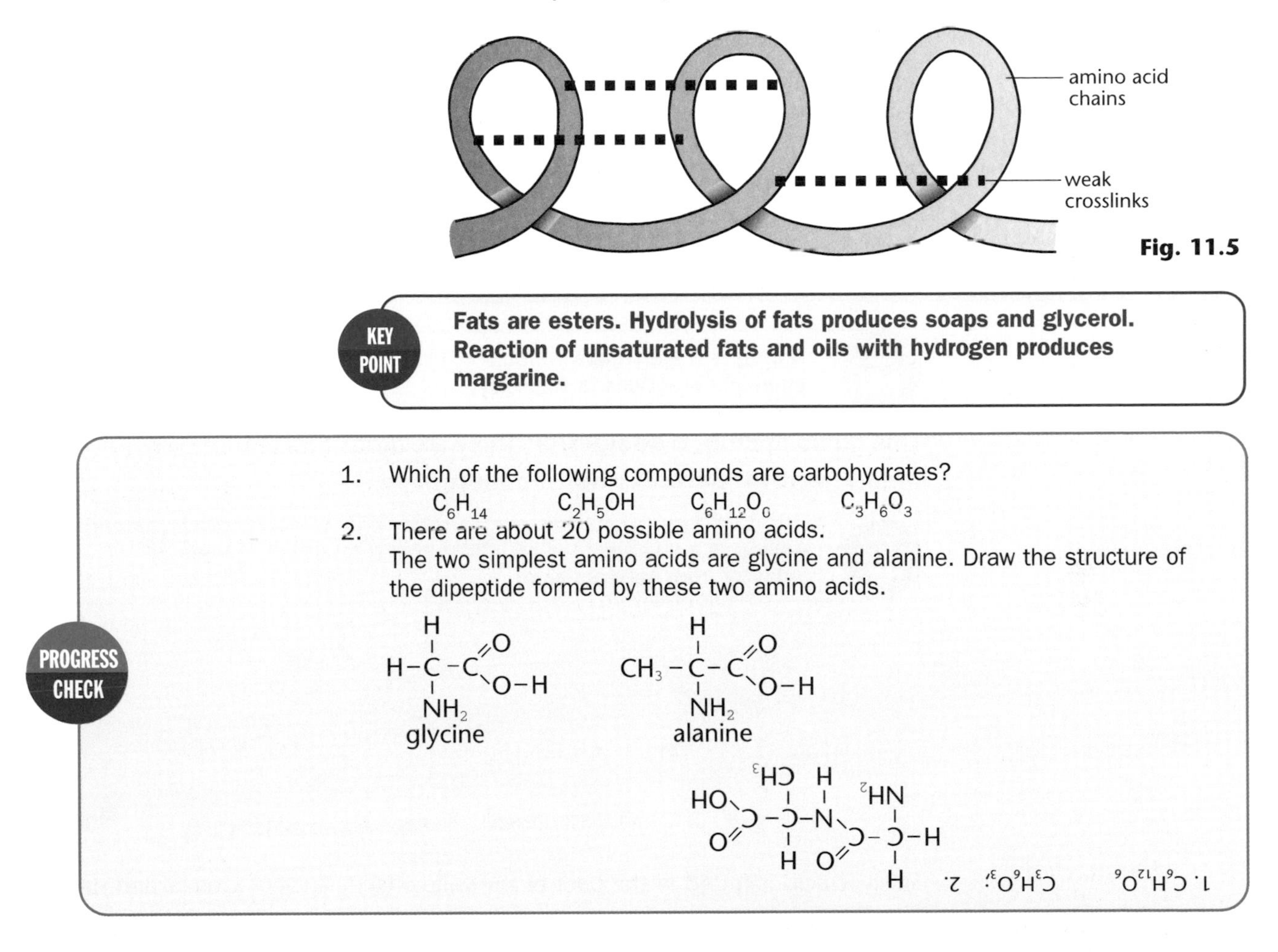

Fig. 11.5

KEY POINT

Fats are esters. Hydrolysis of fats produces soaps and glycerol. Reaction of unsaturated fats and oils with hydrogen produces margarine.

PROGRESS CHECK

1. Which of the following compounds are carbohydrates?
   C_6H_{14} C_2H_5OH $C_6H_{12}O_6$ $C_3H_6O_3$
2. There are about 20 possible amino acids.
   The two simplest amino acids are glycine and alanine. Draw the structure of the dipeptide formed by these two amino acids.

```
     H                        H
     |   //O                  |   //O
 H - C - C             CH3 -  C - C
     |    \O-H                |    \O-H
    NH2                      NH2
  glycine                  alanine
```

1. $C_6H_{12}O_6$, $C_3H_6O_3$; 2.

```
      H   O
      |  //       H
H  -  C - C       |
      |      \N - C - C//O
     NH2      |   |     \OH
              H   CH3
```

Vitamins and food additives

OCR A B

Over a hundred years ago scientists realised that a diet of carbohydrate, proteins and fats is not enough to ensure prolonged good health. Other substances were needed and they were called **vitamins**.

Vitamins can be divided into water-soluble (e.g. vitamin C) and fat-soluble (e.g. vitamin A or vitamin D).
Fat-soluble vitamins can be stored in the body but water-soluble vitamins cannot and have to be taken in on a daily basis. Failure to consume enough vitamins can cause a deficiency disease.

Vitamin C (ascorbic acid) prevents the deficiency disease called scurvy. Vitamin C also destroys harmful free radicals. Fresh fruit and vegetables provide a good source of vitamin C.

Prolonged storage or overcooking can destroy vitamin C.

Raising agents are used when making cakes. **Sodium hydrogencarbonate** is the main ingredient in baking powder. In a cake in the oven the sodium hydrogencarbonate decomposes and bubbles of carbon dioxide are trapped in the cake. This makes the cake rise.

$$2NaHCO_3 \rightarrow Na_2CO_3 + H_2O + CO_2$$

Food additives include artificial colourings, preservatives, emulsifiers and sweeteners.

Paper chromatography can be used to identify which colourings are present.

They:

- have to be approved and are shown with E-numbers, e.g. E101
- have to be included on contents labels
- improve the appearance, keeping properties or flavour of food.

Drugs

OCR A B

A drug is a substance which when taken modifies or affects the chemical reactions in the body.

One group of drugs is **analgesics**. These are drugs that reduce pain. They include **aspirin**, **paracetamol** and **ibuprofen**.

KEY POINT

Aspirin is a chemical called acetyl salicylic acid. It is made in industry from salicylic acid.

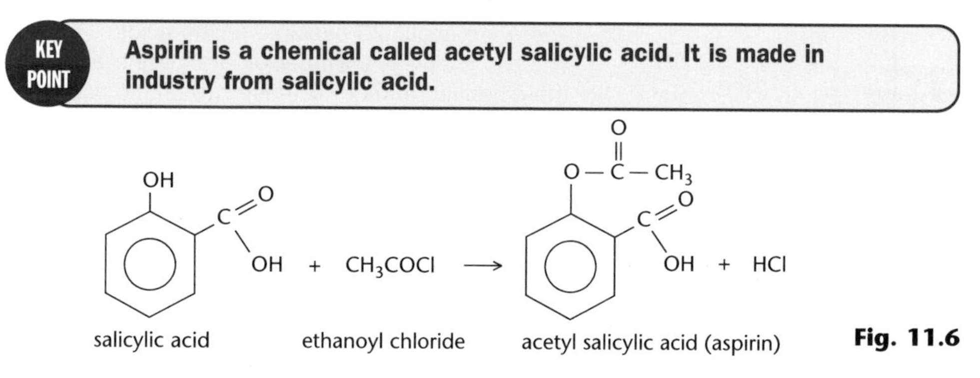

Fig. 11.6

Salicylic acid is found in the bark of the willow tree. Ancient Greeks and native Indians used the bark of willow trees to counter fever and pain. Salicylic acid is bitter and irritates the stomach. In 1890 Felix Hoffman showed how turning salicylic acid into acetyl salicylic acid made it more suitable for use.

Fig. 11.7 below shows the structure of the sodium salt of aspirin. This is more soluble in water as it is an ionic compound.

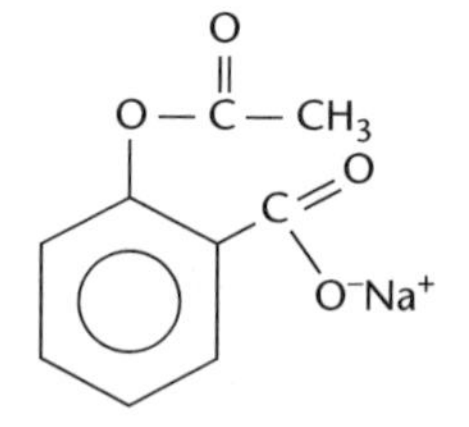

sodium salt of asprin (more soluble)

Fig. 11.7

Sample GCSE question

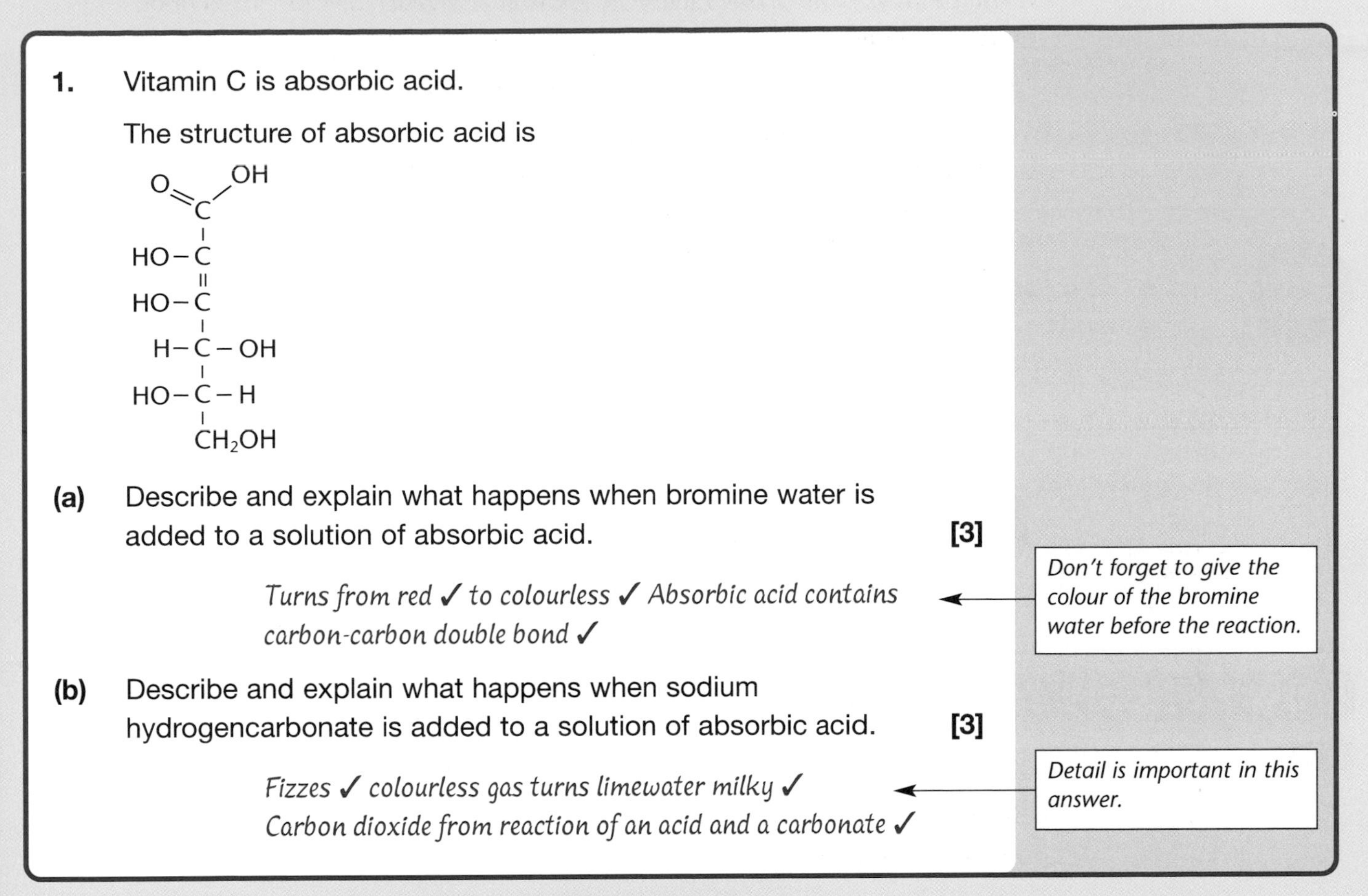

1. Vitamin C is absorbic acid.

The structure of absorbic acid is

```
 O    OH
  \\ /
    C
    |
HO- C
    ||
HO- C
    |
 H- C -OH
    |
HO- C -H
    |
    CH2OH
```

(a) Describe and explain what happens when bromine water is added to a solution of absorbic acid. **[3]**

Turns from red ✓ to colourless ✓ Absorbic acid contains carbon-carbon double bond ✓

Don't forget to give the colour of the bromine water before the reaction.

(b) Describe and explain what happens when sodium hydrogencarbonate is added to a solution of absorbic acid. **[3]**

Fizzes ✓ colourless gas turns limewater milky ✓ Carbon dioxide from reaction of an acid and a carbonate ✓

Detail is important in this answer.

Exam practice question

1. Ammonium hydrogencarbonate, NH_4HCO_3, can be used as a raising agent in biscuits.

(a) Write an equation for the decomposition of ammonium hydrogencarbonate **[2]**

(b) Using sodium hydrogencarbonate can leave a residue in the biscuits. What is the advantage of using ammonium hydrogencarbonate? **[1]**

Aspartame is an artificial sweetener.

Its structure is shown below.

```
             NH2     H   H   H
             |       |   |   |
HOOC - CH2 - C - C - N - C - C - C6H5
             |   ||      |   |
             H   O       |   H
                       COOCH3
```

(c) What is the molecular formula of aspartame? **[1]**

(d) Is aspartame a carbohydrate? Explain your answer **[1]**

(e) Why is aspartame better than sugar for a person on a reduced calorie diet? **[1]**

Chapter 12 Radicals

The following topics are covered in this section:

- ***How radicals form***
- ***Depletion of the ozone layer***

LEARNING SUMMARY

After studying this section you should be able to:

- ***describe how radicals are formed when a covalent bond is broken***
- ***explain how chlorine radicals can deplete the ozone layer.***

KEY POINT

You will probably have heard how holes are appearing in the ozone layer in the upper atmosphere, especially over Antarctica. This could have serious environmental effects. The depletion of this layer has been attributed to the release of substances such as CFCs in aerosols and refrigerators. These can produce radicals that destroy the ozone.

How radicals form

OCR A B

A chlorine molecule contains a **shared pair** of electrons in a **covalent** bond.

$$:\ddot{\underset{..}{Cl}}:\ddot{\underset{..}{Cl}}:$$

If this covalent bond is broken two chlorine radicals are formed. Each of these contains a single unpaired electron.

The chlorine radical (sometimes called a free radical) can be shown as

Only the unpaired electron is shown.

Cl •

Ultraviolet light is of the correct energy to break covalent bonds such as the one between two chlorine atoms.

KEY POINT

Radicals contain an unpaired electron and they have a very short life. They immediately react to form new substances.

PROGRESS CHECK

Hydrogen and chlorine react explosively in sunlight to form hydrogen chloride.

$H_2 + Cl_2 \rightarrow 2HCl$

A few chlorine molecules break up into chlorine radicals.
These chlorine radicals react with hydrogen.

(a) Finish the equation:
Cl• + H-H → H-Cl + ...

(b) The product of this reaction then reacts with another chlorine molecule.
Finish the equation:
...... + Cl—Cl → HCl + ...

(c) Explain why it is necessary to break only a few chlorine molecules into radicals to get a chain reaction.

(d) What is present in sunlight to initiate (start) the process?

(a) H•; (b) H• + Cl—Cl → H—Cl + Cl•; (c) The chlorine radicals are reformed and so can react again; (d) ultraviolet light.

Depletion of the ozone layer

OCR A [B]

Compounds called chlorofluorocarbons (CFCs) were widely used as propellants in aerosols and as refrigerants.

Three elements present in CFCs are chlorine, fluorine and carbon.

CFCs are very **stable** compounds. They are very **unreactive** and do not **burn**. In the stratosphere ultraviolet light breaks down the CFCs forming **chlorine radicals**.

Ozone is a reactive allotrope of oxygen O_3.

The gaseous radicals **catalyse** the breakdown of **gaseous** ozone.

$O_3(g) + Cl\bullet(g) \rightarrow O_2(g) + ClO\bullet(g)$

$O(g) + ClO\bullet(g) \rightarrow O_2(g) + Cl\bullet(g)$

Overall $O_3(g) + O \rightarrow 2O_2(g)$

The O(g) radicals are formed naturally in the stratosphere by the action of ultraviolet light on $O_2(g)$.

100 000 ozone molecules can be broken by one chlorine radical.

KEY POINT **The ozone in the ozone layer removes much of the harmful ultraviolet radiation from the sunlight.**

If more ultraviolet light reaches the surface of the Earth it can cause:

- increased risk of sunburn
- accelerated ageing of the skin
- skin cancer.

Destruction of the ozone layer is a global problem. Why is the Helsinki agreement not enough?

In 1989, 81 nations agreed in Helsinki to ban eight chemicals (mainly CFCs), that damage the ozone layer, by 2001. There is evidence that the holes in the ozone layer, which had been enlarging, could shrink again.

PROGRESS CHECK

1. What radicals can be formed if a hydrogen chloride molecule is broken?
2. What would be formed if a chlorine molecule broke so one atom received both of the pair of electrons?
3. One CFC is CCl_3F. Draw the structural (graphical) formula of this molecule.

Alkanes have been suggested as alternatives to CFCs for aerosols.

4. Suggest one disadvantage of an alkane.
5. A good propellant must be a liquid with a boiling point around room temperature so that it vaporises easily.
   The table gives the boiling points of some alkanes.

Alkane	Boiling point in °C
Methane	–162
Ethane	–89
Propane	–42
Butane	0
Hexane	69

Which one would be best?

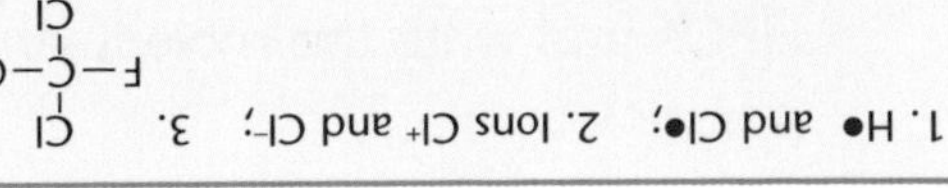

Sample GCSE question

1. Methane, CH_4, reacts with the chlorine in sunlight to produce a mixture of products.

(a) What evidence is there that this reaction involves radicals? **[1]**

Takes place in sunlight ✓

> *Ultraviolet light breaks chlorine–chlorine bonds forming radicals.*

(b) One product of the reaction is chloromethane, CH_3Cl.

(i) Write a balanced equation for the reaction. **[3]**

$CH_4 + Cl_2 \rightarrow CH_3Cl + HCl$ ✓✓✓

(ii) What type of reaction is taking place? **[1]**

Substitution ✓

> *A hydrogen atom is substituted by a chlorine.*

(c) Other products formed include dichloromethane, CH_2Cl_2, and trichloromethane, $CHCl_3$.

Write down the name and formula of another product. **[2]**

CCl_4 ✓ *tetrachloromethane* ✓

Exam practice question

1. There is a number system for CFCs. One CFC is CFC-11. You can work out its formula using the 'rule of 90'.

You add 90 to the number. 90 + 11 = 101

Then the first digit is the number of carbon atoms, the second digit the number of hydrogen atoms and the third digit the number of fluorine atoms. You work out the number of chlorine atoms using a formula

Cl = 2(C+1)-H-F.

In this case, Cl = 2(1+1) – 0 – 1 =3

The formula of the CFC is CCl_3F.

(a) Work out the formula of CFC-12. **[3]**

(b) **(i)** Work out the formula of CFC-132 **[3]**

(ii) CFC-132 can exist as different isomers. Draw the graphical(displayed) formula of two isomers. **[2]**

(c) HFCs are safe alternatives to CFCs. Suggest the three elements combined in HFCs and write the molecular formula of one HFC containing eight atoms per molecule. **[3]**

Chapter 13 Group 2 metals

The following topics are covered in this section:

- ***Physical properties of alkaline earth metals***
- ***Reactions with air and water***
- ***Explaining the difference in reactivity***

LEARNING SUMMARY

After studying this section you should be able to:

- ***compare the physical properties of alkaline earth metals and see trends within the family***
- ***describe the trends in chemical reactions with oxygen and water***
- ***explain the trend in reactivity of the alkaline earth metals.***

KEY POINT

You will know the trend in properties of the metals in group 1 of the periodic table – lithium, sodium, potassium, rubidium and caesium. There are similar trends in properties the elements in group 2 of the periodic table. These elements, called the alkaline earth metals, are beryllium, magnesium, calcium, strontium and barium.

Physical properties of alkaline earth metals

WJEC

The alkaline earth metals are all metals. They conduct electricity.

Table 13.1 below gives the melting points, boiling points and densities of elements in group 2 of the periodic table.

Element	Melting point (°C)	Boiling point (°C)	Density (g/cm^3)
Beryllium	1278	2970	1.85
Magnesium			1.74
Calcium	839	1484	1.54
Strontium	769	1384	2.60
Barium	725	1640	3.51

There is a trend in the melting and boiling points of the metals in group 2 of the periodic table.

Melting and boiling points of group 2 elements are much higher than those of group 1.

KEY POINT

The melting and boiling points of the elements decrease down the group.

There is an exception in the trends. Magnesium seems to have a lower melting point and boiling point than you would expect from the melting and boiling points of the other elements.

KEY POINT

The densities of alkaline earth metals are greater than the densities of alkali metals. None float on water.

PROGRESS CHECK

1. Which of the alkaline earth metals has the highest melting and boiling point?
2. Using the data in the table, what would you expect the melting point of magnesium to be?
3. Using the data in the table, what would you expect the boiling point of magnesium to be?
4. Radium is below barium in group 2. Suggest a melting and boiling point for radium.

1. Beryllium; 2. Around 1000°C; 3. Around 2000°C; 4. Around 700°C and around 1200°C

Reactions with air and water

WJEC

The alkaline earth metals are less reactive than the alkali metals in group 1 of the periodic table.

There is a trend in the reactivity of the elements.

Magnesium also reacts with nitrogen to form magnesium nitride. Lithium behaves in a similar way.

Reactions with air

Magnesium burns brightly in air or oxygen to form **magnesium oxide**.

$2Mg(s) + O_2(g) \rightarrow 2MgO(s)$

Magnesium oxide is a white solid. A damp piece of universal indicator paper turns purple showing an alkaline oxide has been formed.

Calcium burns in air or oxygen to form **calcium oxide**.

$2Ca(s) + O_2(g) \rightarrow 2CaO(s)$

Reactions with water

Magnesium hydroxide would be formed but at the high temperature it decomposes into magnesium oxide and steam.

Magnesium hardly reacts with cold water and reacts very slowly with boiling water.

It does react rapidly with steam to produce **magnesium oxide** and **hydrogen**.

$Mg(s) + H_2O(g) \rightarrow MgO(s) + H_2(g)$

Calcium reacts slowly with cold water to form **calcium hydroxide** and **hydrogen**.

$Ca(s) + 2H_2O(l) \rightarrow Ca(OH)_2(aq) + H_2(g)$

Barium reacts steadily with cold water to form **barium hydroxide** and **hydrogen**.

$Ba(s) + 2H_2O(l) \rightarrow Ba(OH)_2(aq) + H_2(g)$

There is an increase in reactivity down the group.

Beryllium – least reactive

Magnesium

Calcium

Strontium

Barium – most reactive

Group 2 metals are generally less reactive than the metals in group 1.

The metals in both groups 1 and 2 increase in reactivity down the group.

PROGRESS CHECK

1. One alkaline earth metal does not react with cold water. Which metal is this?
2. What is the product of the reaction between strontium, Sr, and oxygen?
3. Write a symbol equation for the reaction of strontium and oxygen.
4. What are the products of the reaction between strontium and cold water?
5. Write a symbol equation for the reaction of strontium and water.
6. One of the alkaline earth metals is normally stored under paraffin oil like alkali metals. Which metal is this? Explain your choice.

1. Beryllium; 2. Strontium oxide; 3. $2Sr(s) + O_2(g) \rightarrow 2SrO(s)$; 4. Strontium hydroxide and hydrogen; 5. $Sr(s) + 2H_2O(l) \rightarrow Sr(OH)_2(aq) + H_2(g)$; 6. Barium. It is an extremely reactive alkaline earth metal.

Explaining the difference in reactivity

WJEC

Table 13.2 below gives the atomic radii and the electron arrangement of atoms of the elements in group 2 of the periodic table.

Element	Atomic radius in arbitrary units	Electron arrangement
Beryllium	112	2,2
Magnesium	160	2,8,2
Calcium	197	2,8,8,2
Strontium	215	2,8,18,8,2
Barium	217	2,8,18,18,8,2

Down group 2 the atoms increase in size. This is a similar trend in group 1

When alkaline earth metals react , each atom loses two electrons to form an ion with a 2+ charge.

e.g. $Mg \rightarrow Mg^{2+} + 2e^-$

Alkaline earth metals (group 2) are less reactive than alkali metals (group 1) because more energy is required to lose two electrons than to lose one electron.

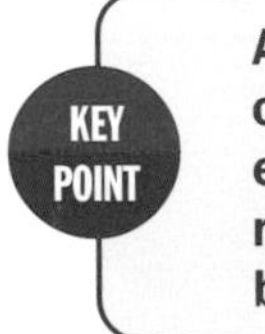

All elements in group 2 have atoms containing two electrons in the outer shell. These two electrons are further from the nucleus. These electrons are lost more easily the further they are away from the nucleus. This is because there is a reducing force of attraction between the nucleus and the outer electrons.

PROGRESS CHECK

1. Which is the least reactive alkaline earth metal?
2. Write an ionic equation for the change when a strontium atom reacts.
3. The table compares some differences between a magnesium atom and a barium atom.

	Magnesium atom	Barium atom
No. of protons in nucleus	12	56
Distance of outer electrons from nucleus in arbitrary units	160	217
Number of electron shells	3	6

Which one of the differences does not explain the increased reactivity of barium?

Radium is another element in group 2.

4. How does radium compare in reactivity with the other elements in group 2?

1. Beryllium; 2 $Sr \rightarrow Sr^{2+} + 2e^-$; 3. No. of protons in the nucleus; 4. Most reactive.

Sample GCSE question

1. When Dalton listed the elements in 1803 he believed that lime was an element.

Calcium was isolated as an element in 1898 by Sir Humphrey Davy. He was able to use electricity to extract calcium from calcium chloride.

(a) Suggest why Dalton believed that lime was an element. **[2]**

Lime is very difficult to split up ✓.
Elements are substances that cannot be split up ✓.

This is an Ideas and Evidence question.

(b) Outline how Davy extracted calcium from calcium chloride. **[3]**

Electrolysis ✓ of molten calcium chloride ✓.
Calcium is formed at the cathode (negative electrode) ✓.

(c) Davy added a small piece of calcium to cold water.

(i) Describe what he saw. **[4]**

The calcium sank ✓. It fizzed as it reacted with water ✓.
A colourless gas was produced ✓. A colourless but slightly cloudy solution was formed ✓.

Calcium sinks because it has a greater density than water i.e. 1 g/dm³.

(ii) Write a balanced symbol equation for the reaction. **[3]**

$Ca + 2H_2O \rightarrow Ca(OH)_2 + H_2$ ✓✓✓

One mark for correct formulae of reactants, one for correct formulae of products and one mark for correct balancing.

Exam practice question

1. (a) How do group 2 metals compare in reactivity with group 1 metals? Explain this in terms of the structure of the atoms. **[3]**

(b) How does the reactivity of group 2 metals change down the group? Explain this in terms of the structure of the atoms. **[3]**

Exam practice answers

Chapter 1 Classifying materials

1 (a) (i)

One mark for the pair of electrons between the two atoms.
One mark for other electrons correctly shown. **[2]**
(ii) Covalent bond **[1]**

(b) Covalent bond breaks
Forms two ions
H^+ and Cl^-
Electricity is transferred by ions.
Any three points. **[3]**

2 Substance A has a giant structure of ions .
Giant structure because of high melting point and conducts electricity when molten but not when solid.
Substance B has a metallic structure.
Conducts electricity when solid.
Substance C has a molecular structure.
Low melting and boiling point.
Substance D has a giant structure of atoms.
High melting point and does not conduct electricity. **[8]**

Chapter 2 Changing materials

1 (a) Graph fills over half the grid and labelled axes. **[1]**
Correct plotting **[1]**
Curve drawn **[1]**
(b) At the start of the reaction. **[1]**
(c) 40s **[1]**
(d) When 0.05g had reacted, 50 cm^3 of gas had been given off; from the graph this is after 18s. **[1]**
(e) The graph is steeper **[1]**
Reaches the same final level **[1]**
Powder has a larger surface area **[1]**
Same mass of magnesium used. **[1]**

2 (a) $3CuBr + Fe \rightarrow 3Cu + FeBr_3$ **[3]**
Formulae on LHS **[1]**
Formula on RHS **[1]**
Balancing **[1]**
(b) Displacement reaction1
(c) Add mixture to dilute hydrochloric acid **[1]**
All substances dissolve except copper **[1]**
Filter off copper, wash and dry **[1]**
(d) 144g of copper bromide **[1]**
produces 64g of copper **[1]**
9.6g of copper **[1]**

Chapter 3 Patterns of behaviour

1 (a) (i) Reactivity increases down the group. **[1]**
(ii) Atoms of all elements have two electrons in the outer shell **[1]**
As the group is descended these two electrons are further from the nucleus **[1]**
Force of attraction between nucleus and electrons is weaker. **[1]**
Electrons are more easily lost. **[1]**
(b) (i) $CaCl_2$ **[1]**
(ii) Add dilute hydrochloric acid **[1]**
To a measured amount of calcium hydroxide with indicator **[1]**
Until indicator changes colour. **[1]**
Repeat without indicator. **[1]**
Evaporate until small volume of solution remains. **[1]**
Leave to cool and crystallise. **[1]**
Any five points

2

New condition	Change, if any	Explanation
Use 5 g of powdered zinc	Faster **[1]**	Larger surface area **[1]**
Use 40 °C	Faster **[1]**	Higher temperature, particles move faster – more collisions **[1]**
Use 100 cm^3 of hydrochloric acid (50 g/dm^3)	Slower **[1]**	Lower concentration – fewer collisions between acid particles and zinc **[1]**
Use 100 cm^3 of ethanoic acid (100 g/dm^3)	Slower **[1]**	Ethanoic acid is a weak acid – only partially ionised **[1]**

Chapter 4 Water

1 (a) Temperature on x-axis and correct scales on x and y axis **[1]**
Correct plotting **[1]**
Curve drawn **[1]**
(b) (i) 43–48g **[1]**
(ii) 43–45g **[1]**
(c) (i) 25–27°C **[1]**
(ii) Crystals form **[1]**
19g of crystals formed **[1]**

2 (a) White **[1]** to blue **[1]**
(b) Boils at 102°C or white residue of evaporation – aqueous solution. **[1]** pH7 – pure water **[1]**

Chapter 5 Acids, bases and salts

1 This is a sample answer. There would be other ways of doing it.
(a) Add dilute hydrochloric acid and barium chloride to each solution **[1]**
Sodium sulphate and sulphuric acid produce white precipitates. **[1]**
Sodium chloride does not form a precipitate and can be identified. **[1]**
Add litmus to samples of sodium sulphate and sulphuric acid. **[1]**

Sulphuric acid turns litmus red **[1]**
Sodium sulphate turns litmus purple. **[1]**
(b) $Ba^{2+}(aq) + SO_4^{2-}(aq) \rightarrow BaSO_4(s)$ **[2]**

2 Use solution of soluble lead (II) salt **[1]**
e.g. lead (II) nitrate. **[1]**
Mix solutions **[1]**
Filter off lead (II) chromate **[1]**
Wash with distilled water **[1]**
1 mark for quality of written communication.
$Pb(NO_3)_2 + K_2CrO_4 \rightarrow PbCrO_4 + 2KNO_3$ **[3]**

Chapter 6 Metals and redox

1 Name of iron ore e.g. haematite
Iron ore, limestone and coke added to the blast furnace.
Furnace heated with hot air.
Iron tapped off the bottom of the furnace.
This is pig iron or contains unwanted impurities
Pig iron (and scrap iron) and limestone added
Heated until molten.
Oxygen (or air) blown onto the surface of the molten iron.
Oxidises impurities
Other impurities removed as slag.
The required amounts of carbon added.
Seven marks for seven points above. 1 for QWC

Chapter 7 Further Carbon Chemistry

1 (a) (i) carbon, hydrogen and oxygen – 3 correct **[2]**; 2 correct **[1]**.
(ii) Fermentation **[1]**
(iii) Yeast or enzymes or zymase **[1]**
(b) Fractional distillation of crude oil **[1]**
Cracking (or thermal decomposition) **[1]**
Of high boiling point fractions **[1]**
Using high temperature and catalyst **[1]**
(c) making poly(ethene) **[1]**
(d) ethene + water → ethanol **[1]**
$C_2H_4 + H_2O \rightarrow C_2H_5OH$ **[2]**
(e) (i) Oxygen is added. **[1]**
(ii) Cold air enters the apparatus and warm air leaves. **[1]**

Chapter 8 Further Quantitative Chemistry

1 (a) (i) Pipette **[1]**
(ii) Burette **[1]**
(b) An indicator **[1]**
e.g. phenolphthalein ,which changes from colourless to pink. **[1]**
(c) (i) $35/1000 \times 1.2 = 0.042$ **[1]**
(ii) $0.5 \times 0.042 = 0.021$ **[1]**
(iii) $1000/25 \times 0.021 = 0.84$ **[1]**

2 (a) $\frac{20 \times 0.1}{1000}$ **[1]** = 0.002 **[1]**
(b) 0.002 **[1]**
Note: The equation shows 1 mole of ethanoic acid reacts with 1 mole of sodium hydroxide.
(c) 0.04 moles **[1]**
(d) 60g **[1]**
(e) 0.04 x 60 **[1]** = $2.40 g/dm^3$ **[1]**

Chapter 9 Electrochemistry and Electrolysis

(a) Cl^-; chlorine, H^+; hydrogen; OH^-; oxygen; H^+; hydrogen; OH^-; oxygen; Cu^{2+}; Cu. Half mark for each correct answer – rounded up to whole mark. **[6]**
(b) At the anode (positive electrode) **[1]**
Copper goes into solution **[1]**
As copper(II) ions **[1]**

Chapter 10 Collection of gases

1 (a) $NH_4Cl + NaOH \rightarrow NH_3 + NaCl + H_2O$ **[1]**
(b) (i) Ammonia dissolves in cold water **[1]**
(ii) Use damp red litmus **[1]**
Hold it near the mouth of the test tube **[1]**
Test tube full of ammonia when the litmus turns blue **[1]**

Chapter 11 Food and drugs

1 (a) $NH_4HCO_3(s) \rightarrow NH_3(g) + H_2O(g) + CO_2(g)$ **[2]**
(b) When ammonium hydrogencarbonate decomposes it leaves behind no solid residue. **[1]**
(c) $C_{14}H_{18}O_5N_2$ **[1]**
(d) Not a carbohydrate. Does not fit formula for carbohydrate. **[1]**
(e) Does not have the high energy content of a carbohydrate. **[1]**

Chapter 12 Radicals

1 (a) CF_2Cl_2 **[3]**
(b) (i) $C_2H_2F_2Cl_2$ **[3]**
(ii)

```
    Cl  Cl            F   H
    |   |             |   |
F - C - C - F     F - C - C - H
    |   |             |   |
    H   H             Cl  Cl
```

[2]

There are other possibilities
(c) Hydrogen, fluorine and carbon, three correct – **[2]**, two correct **[1]**
e.g. $C_2H_2F_4$ **[1]**

Chapter 13 Group 2 Metals

1 (a) Group 2 metals are less reactive than corresponding group 1 metals. **[1]**
Group 2 metals have 2 electrons in outer shell and group 1 have 1 electron. **[1]**
More difficult to lose 2 electrons than 1. **[1]**
(b) Reactivity increases down group. **[1]**
Atoms large down the group. **[1]**
Easier to lose electrons further from the nucleus. **[1]**

Centre number
Candidate number
Surname and initials

Letts Examining Group

General Certificate of Secondary Education

Chemistry
Higher Tier
Paper 1

Time: one and a half hours

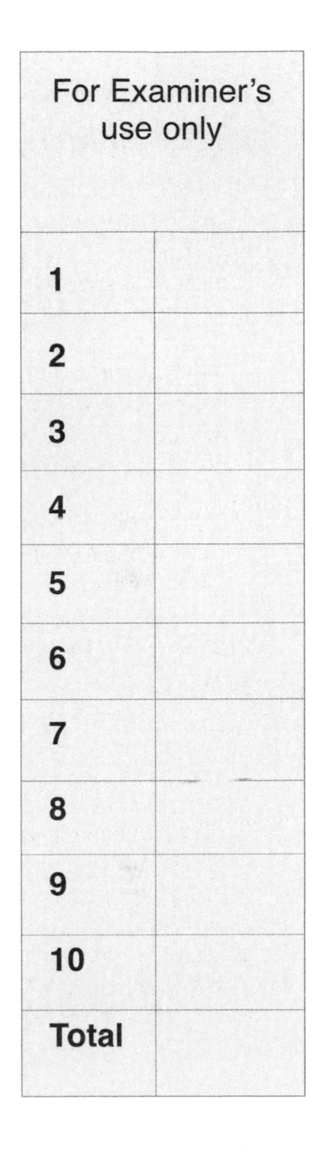

For Examiner's use only	
1	
2	
3	
4	
5	
6	
7	
8	
9	
10	
Total	

Instructions to candidates

Write your name, centre number and candidate number in the boxes at the top of this page.

Answer ALL questions in the spaces provided on the question paper.

Show all stages in any calculations and state the units.
You may use a calculator.

Include diagrams in your answers where this may be helpful.

Information for candidates

The maximum mark for this paper is 100.

The number of marks available is given in brackets **[2]** at the end of each question or part question.

The marks allocated and the spaces provided for your answers are a good indication of the length of answer required.

Where you see this icon you will be awarded marks for the quality of written communication in your answers.
This means, for example, that you should:

- write in sentences
- use correct spelling, punctuation and grammar
- use correct scientific terms.

Leave blank

1 Copper(II) oxide is added to dilute sulphuric acid in a test tube and the mixture warmed. When the test tube is left in a rack for a few minutes a black solid settles to the bottom, leaving a clear blue liquid.

(a) **(i)** Write down the name of the clear blue liquid.

.... **[1]**

(ii) Write down the name of the black solid in the test tube.

.... **[1]**

(iii) Write a balanced equation for the reaction between copper(II) oxide and sulphuric acid.

.... **[2]**

(iv) What name is given to this type of reaction?

.... **[1]**

(b) Describe how you would make blue crystals from the mixture in the test tube. (One mark is for the correct sequencing of your answer.)

....

....

....

.... **[3+1]**

(c) The reaction between copper(II) oxide and sulphuric acid gives out energy in the form of heat.

(i) What name is given to a reaction which gives out heat?

.... **[1]**

(ii) Draw an energy level diagram to represent this reaction.

[3]

(Total 13 marks)

Leave blank

2 The apparatus shown in the diagram was used to study the combustion of a liquid hydrocarbon, octane, C_8H_{18}.

Gases from the burning hydrocarbon were drawn through the apparatus for several minutes.

(a) A clear, colourless liquid appeared at A.

(i) Name this liquid.

.... **[1]**

(ii) Describe a test to prove the identity of this liquid.

....

.... **[2]**

(b) (i) What would you see at B as the experiment was carried out?

....

.... **[2]**

(ii) What does this show about the gases produced by combustion of the hydrocarbon?

.... **[1]**

(c) (i) Write a balanced equation for the complete combustion of octane.

.... **[2]**

(ii) When the experiment was completed, a black deposit of carbon was noted at C.
Explain how this was formed.

....

.... **[2]**

(Total 9 marks)

3 Sarah studied the reaction between hydrochloric acid and sodium carbonate.

Leave blank

$$2HCl + Na_2CO_3 \rightarrow 2NaCl + CO_2 + H_2O$$

She made hydrochloric acid of different concentrations by mixing a more concentrated solution with water. She used tablets each of which contained the same mass of sodium carbonate. She timed how long it took for a tablet to react completely in the same volume of each concentration of acid.
Her results are shown in the table.

volume of acid in cm^3	volume of water in cm^3	time for tablet to react in seconds
2	18	350
4	16	245
6	14	220
8	12	142
10	10	57

(a) Plot the volume of acid used against time on the grid below.
Draw the line of best fit for the points you have plotted. **[3]**

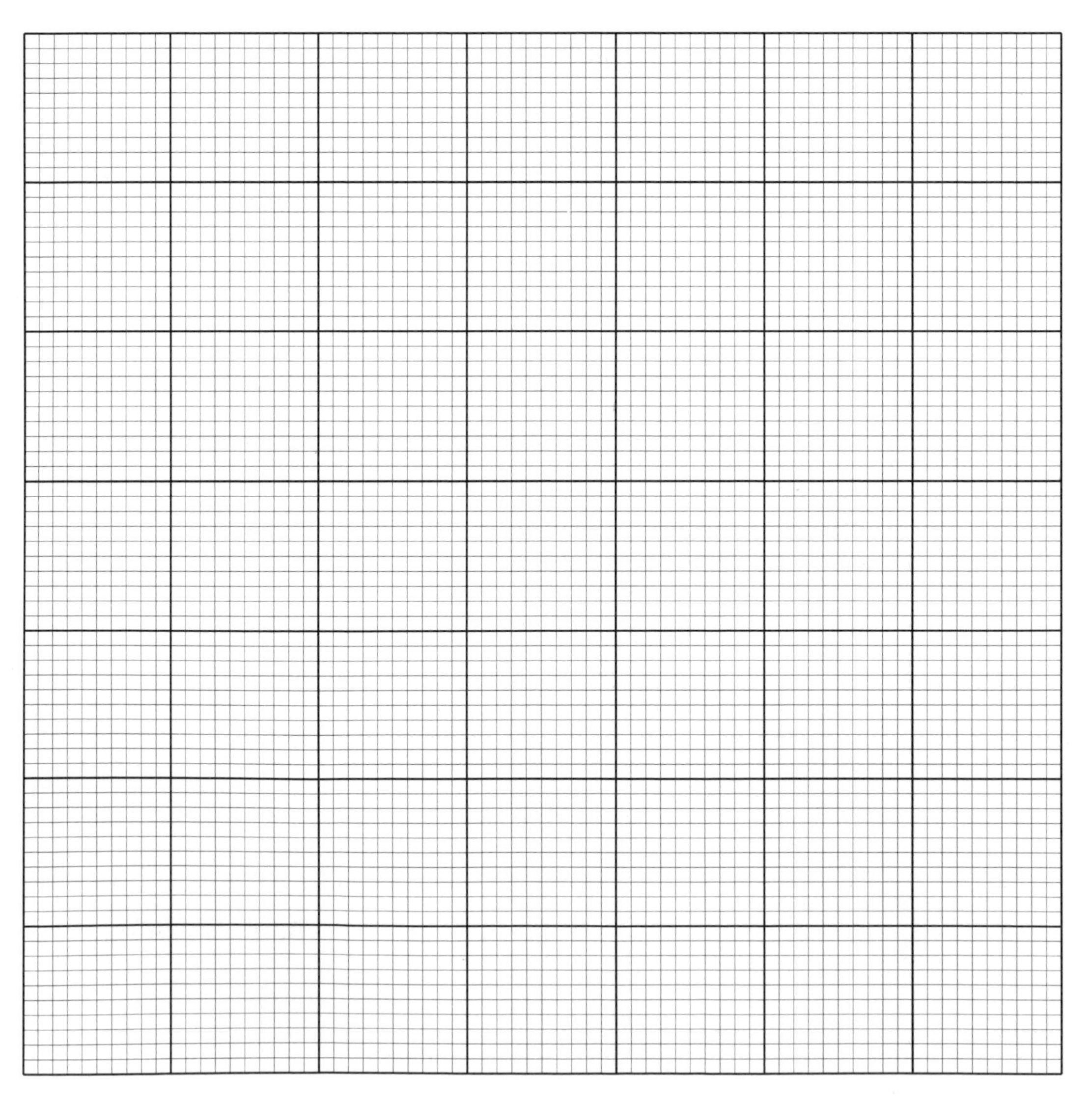

Leave blank

(b) Sarah was careful to ensure fair testing in her experiments.

(i) Explain how mixing each volume of acid with a different volume of water helped to ensure fair testing.

....

....

.... **[2]**

(ii) What other thing, not mentioned above, must Sarah have kept constant to ensure fair testing?

.... **[1]**

(c) One of Sarah's results is anomalous.

(i) What volume of acid was used for the anomalous result?

.... **[1]**

(ii) Suggest what may have caused the error in this result.

....

.... **[1]**

(d) (i) Describe the relationship between concentration of acid and rate of this reaction shown by Sarah's results.

....

....

.... **[2]**

(ii) Use your knowledge of particles to explain this relationship.

....

....

....

.... **[3]**

(Total 13 marks)

4 The apparatus shown below was used to find the formula of an oxide of copper.

Leave blank

excess hydrogen burnt off

oxide of copper

hydrogen

HEAT

Hydrogen gas was passed over a heated sample of the oxide of copper in a ceramic container. The oxide of copper was reduced to copper metal. The following masses were measured.

mass of ceramic container = 12.64 g
mass of ceramic container + oxide of copper = 15.88 g
mass of ceramic container + copper = 15.52 g

(a) (i) What mass of the copper was formed in the experiment?

.... **[1]**

(ii) What mass of oxygen was combined with this mass of copper?

.... **[1]**

(b) Use the values from (a) to work out the formula of this oxide of copper. (Relative atomic masses: Cu = 64, O = 16)

[4]

(c) This method could not be used to find the formula of sodium oxide. Explain why.

....

.... **[2]**

(Total 8 marks)

Leave blank

5 The apparatus below was used to break the large hydrocarbon molecules in petroleum jelly into smaller hydrocarbon molecules.

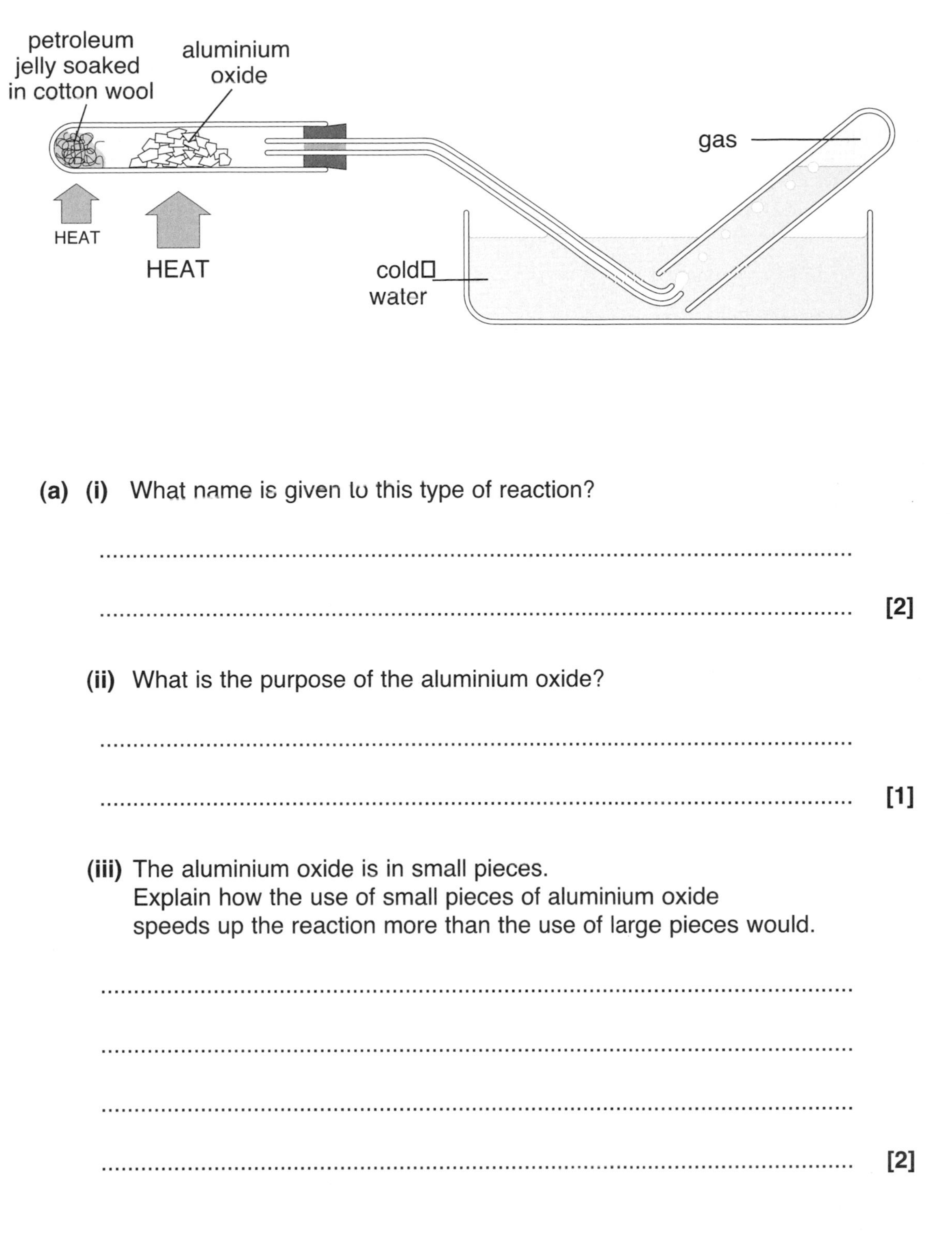

(a) **(i)** What name is given to this type of reaction?

....

.... **[2]**

(ii) What is the purpose of the aluminium oxide?

....

.... **[1]**

(iii) The aluminium oxide is in small pieces.
Explain how the use of small pieces of aluminium oxide speeds up the reaction more than the use of large pieces would.

....

....

....

.... **[2]**

Leave blank

(b) The gas collected in the tube has a molecule containing two carbon atoms.

(i) Name the gas collected in the tube.

.... **[1]**

(ii) Describe how you would test this gas to show that it is not an alkane.

....

....

.... **[2]**

(c) The alkane decane, $C_{10}H_{22}$, can be broken down in a similar reaction to give octane, C_8H_{18}, and the gas in (b)(i).
Draw graphical (displayed) formulae to show the equation for this reaction.

[3]

(Total 11 marks)

Leave blank

6 This diagram shows a cross section through some layers of rock in the Earth's crust.

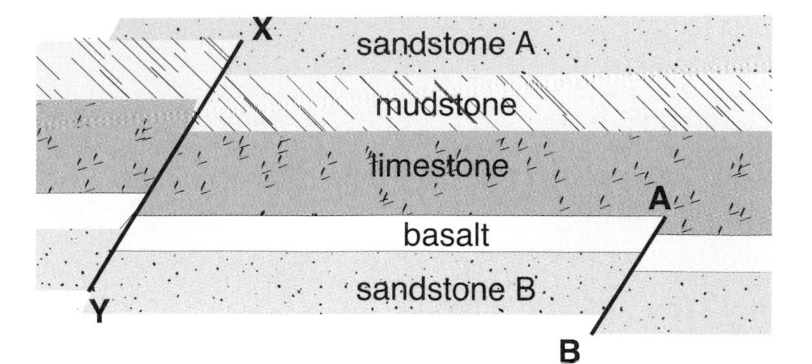

(a) Which of the rocks shown in the diagram:

(i) is the youngest sedimentary rock? .... **[1]**

(ii) is an igneous rock? .... **[1]**

(iii) is mostly calcium carbonate? .... **[1]**

(b) The basalt is made up from small crystals.
Explain what this tells you about how it was formed.

....

....

.... **[2]**

(c) How can you tell that the fault A–B occurred before the fault X–Y?

....

....

.... **[2]**

(Total 7 marks)

7 The diagram shows the carbon cycle.

Leave blank

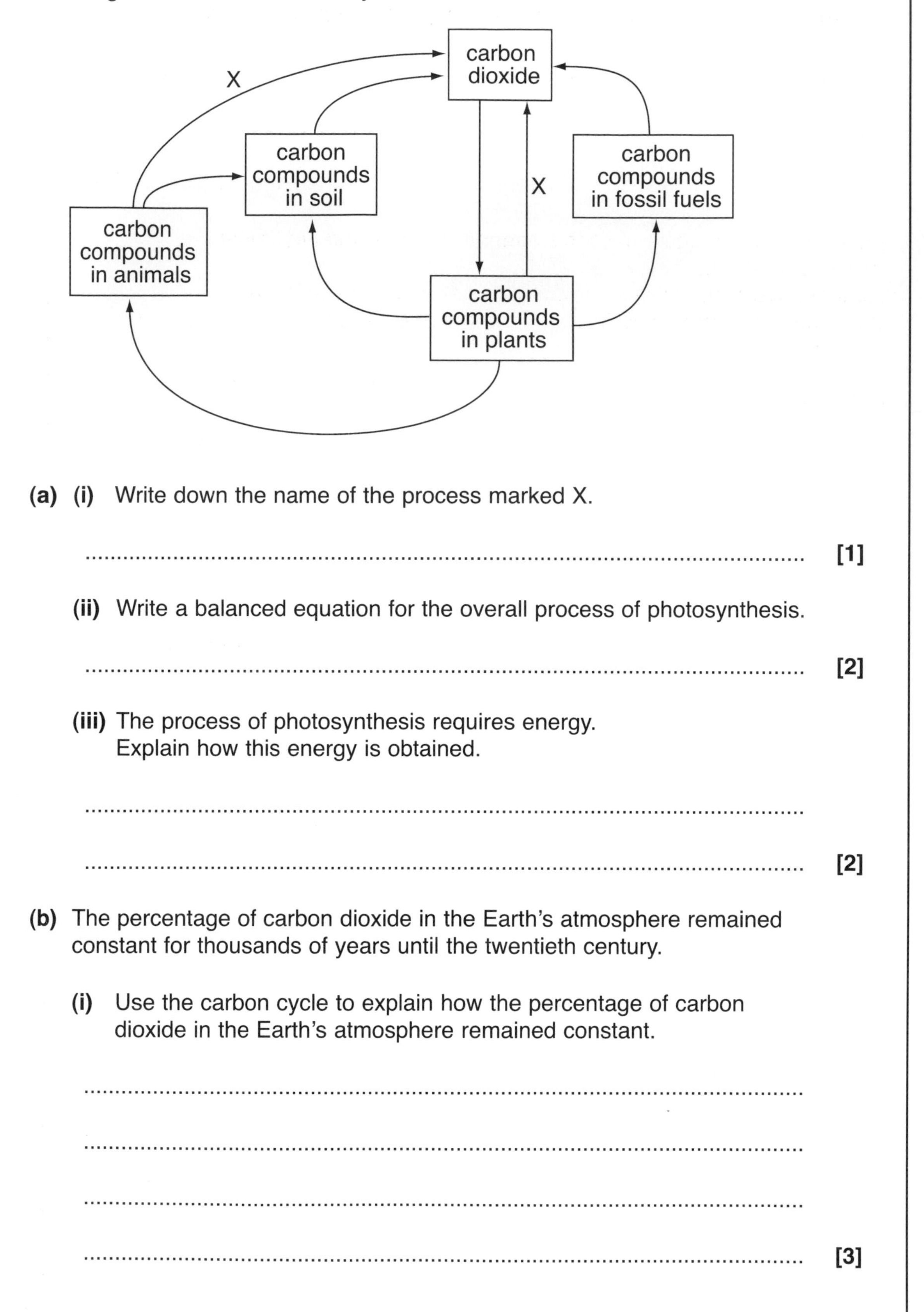

(a) **(i)** Write down the name of the process marked X.

.... **[1]**

(ii) Write a balanced equation for the overall process of photosynthesis.

.... **[2]**

(iii) The process of photosynthesis requires energy.
Explain how this energy is obtained.

....

.... **[2]**

(b) The percentage of carbon dioxide in the Earth's atmosphere remained constant for thousands of years until the twentieth century.

(i) Use the carbon cycle to explain how the percentage of carbon dioxide in the Earth's atmosphere remained constant.

....

....

....

.... **[3]**

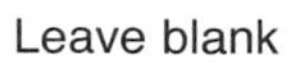

(ii) Describe and explain the change in percentage of carbon dioxide in the Earth's atmosphere which occurred during the twentieth century.

....

....

.... **[2]**

(c) Four billion years ago the Earth's atmosphere contained hydrogen, helium and ammonia.
Suggest how these gases were removed from the atmosphere.
(One mark for correct spelling, punctuation and grammar.)

....

....

....

....

.... **[4+1]**

(Total 15 marks)

Leave blank

8 Ammonia is manufactured from nitrogen and hydrogen.

$$N_2(g) + 3H_2(g) \rightleftharpoons 2NH_3(g)$$

(a) What does the symbol $\rightleftharpoons$ show?

.... **[1]**

(b) Where are the raw materials nitrogen and hydrogen obtained from?

nitrogen is obtained from ....

hydrogen is obtained from .... **[2]**

(c) Suggest two reasons why the manufacture of ammonia is carried out at high pressure.

1 ....

2 .... **[2]**

(d) At low temperatures a very high yield of ammonia can be obtained if the mixture is left long enough. Explain why the process is actually carried out at higher temperatures which give a lower yield.
Use ideas about collisions between particles in your answer.

....

....

.... **[3]**

(e) Why was the discovery of a method to make ammonia on a large scale globally important?

....

....

.... **[2]**

(f) Calculate the mass of ammonia that could be produced if 28 tonnes of nitrogen is completely converted into ammonia.
(Relative atomic masses: H = 1, N = 14.)

mass = ... tonnes **[2]**

(Total 12 marks)

9 The diagram below shows part of the Periodic Table.

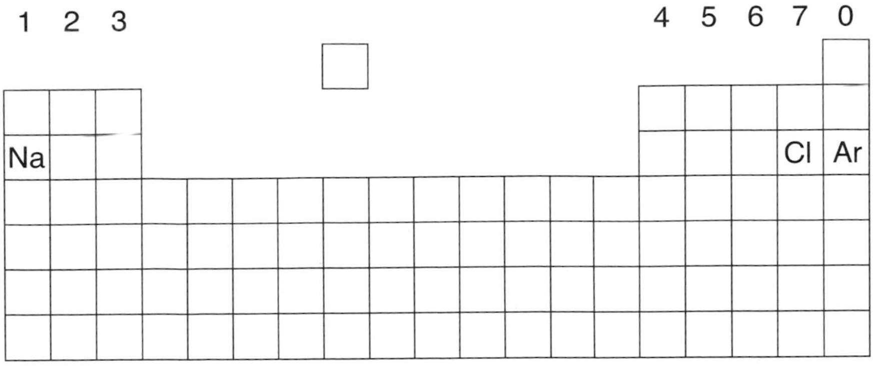

1 2 3 4 5 6 7 0

Na Cl Ar

The position of three elements in the Periodic Table is shown.

(a) Describe the difference in the atomic structure of these three elements.

[2]

(b) Using these three elements as examples, describe the trend in chemical properties across the second period of the Periodic Table.

[3]

(c) Use ideas about the electronic structure of the three elements to explain this trend in chemical properties.

[2]

(Total 7 marks)

Leave blank

10 The table shows the composition of the ocean.

Leave blank

Ion	Concentration of ion in g/100g of sea water
Chloride Cl^-	19.2
Sodium Na^+	10.7
Sulphate SO_4^{2-}	2.7
Magnesium Mg^{2+}	1.4
Calcium Ca^{2+}	0.4
Potassium K^+	0.38
Hydrogencarbonate HCO_3^-	0.14
Bromide Br^-	0.07

Rocks such as halite (sodium chloride) and calcite (calcium sulphate) are dissolved in rivers and enter the oceans. Other rocks such as limestone, chalk or marble react with rain water and get washed into rivers.

(a) Suggest why there are low concentrations of calcium ions in the ocean despite large quantities of calcium compounds in river water.

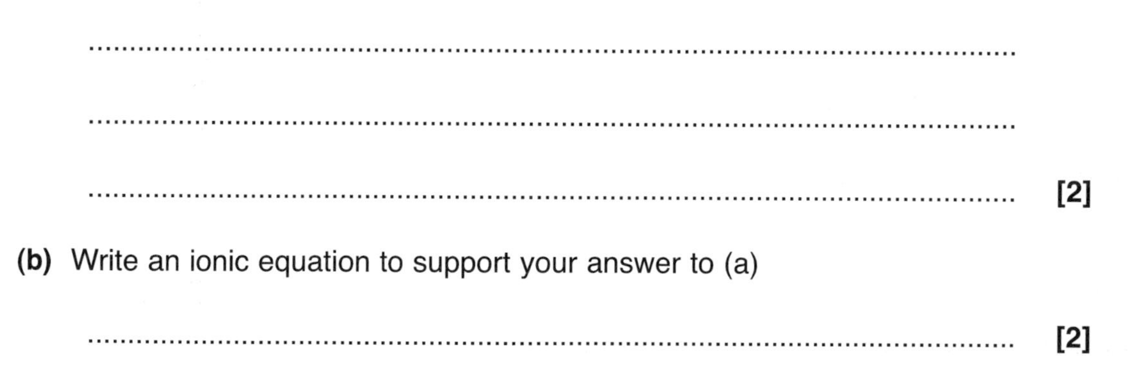

....

....

.... **[2]**

(b) Write an ionic equation to support your answer to (a)

.... **[2]**

(Total 4 marks)

Centre number
Candidate number
Surname and initials

Letts Examining Group

General Certificate of Secondary Education

Chemistry
Higher Tier
Paper 2

Time: one hour

Instructions to candidates

Write your name, centre number and candidate number in the boxes at the top of this page.

Answer ALL questions in the spaces provided on the question paper.

Show all stages in any calculations and state the units.
You may use a calculator.

Include diagrams in your answers where this may be helpful.

Information for candidates

The maximum mark for this paper is 60.

The number of marks available is given in brackets **[2]** at the end of each question or part question.

The marks allocated and the spaces provided for your answers are a good indication of the length of answer required.

Where you see this icon you will be awarded marks for the quality of written communication in your answers.
This means, for example, that you should:

- write in sentences
- use correct spelling, punctuation and grammar
- use correct scientific terms.

For Examiner's use only	
1	
2	
3	
4	
5	
6	
Total	

Leave blank

1 The table gives some information about the homologous series of alkanes.

name	formula	relative molecular mass	boiling point in°C
methane	CH_4	16	–161
ethane	C_2H_6	30	....
....	C_3H_8	44	–42
butane	....	58	–1
pentane	C_5H_{12}	72	36

(a) Complete the table by filling in the three blank boxes. **[3]**

(b) Explain what is meant by the term homologous series as it applies to the alkanes.

....

....

.... **[2]**

(c) Butane exists as a number of structural isomers.

(i) What are structural isomers?

....

.... **[2]**

(ii) Draw structural (displayed) formulae for **two** structural isomers of butane.

[2]

(Total 9 marks)

Leave blank

2 Simon investigated the solubility of potassium nitrate in water.
He measured the mass of potassium nitrate which would dissolve in $100\,cm^3$ of water at different temperatures.
His results are shown in the table.

temperature in °C	solubility of potassium nitrate in g per $100\,cm^3$ water
20	32
30	47
40	63
50	65
60	110
70	138

(a) Plot the solubility of potassium nitrate against temperature on the grid below. Draw the line of best fit for the points you have plotted. **[3]**

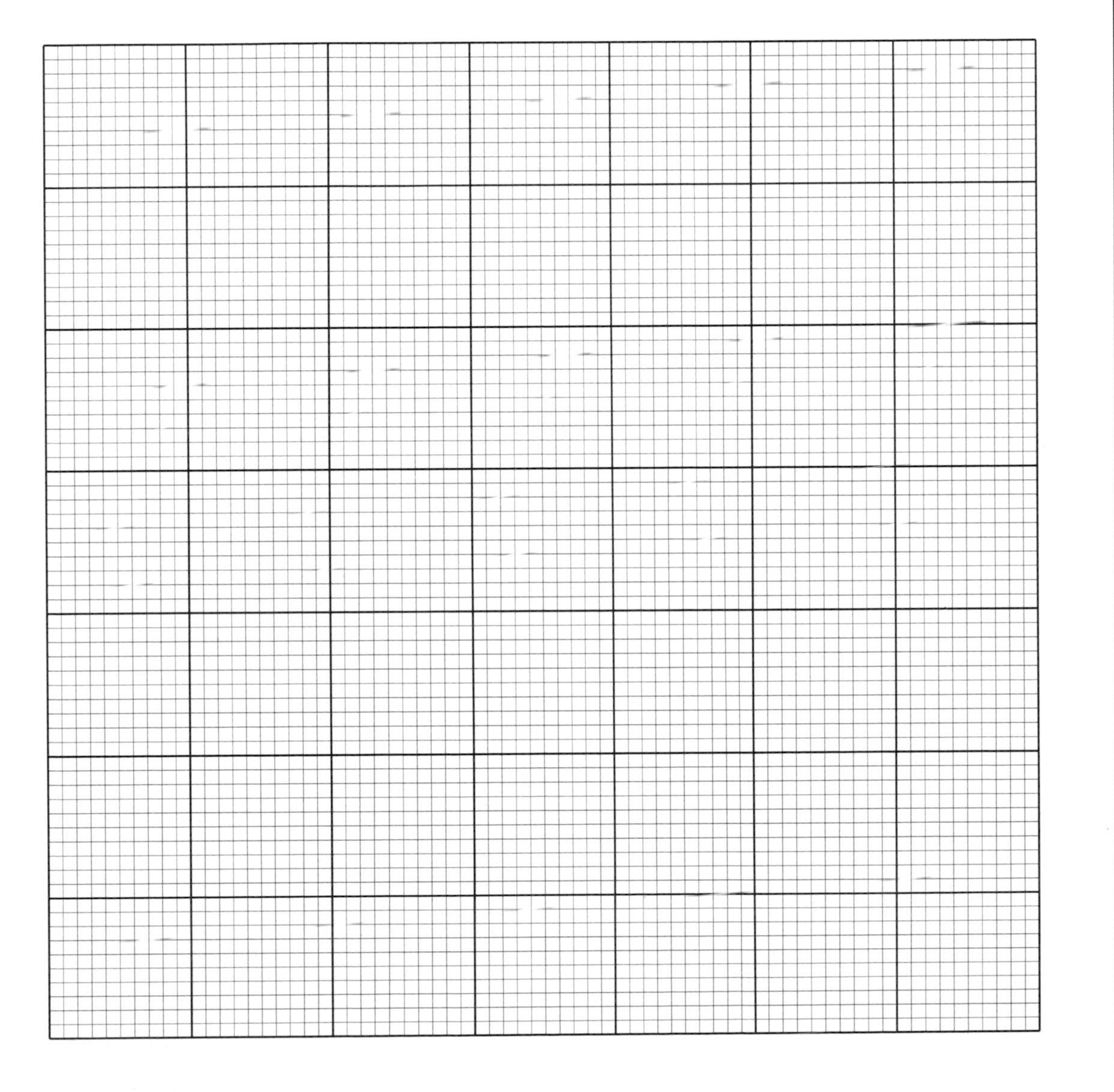

Leave blank

(b) Describe the way in which the solubility of potassium nitrate changes with temperature.

....

....

.... **[2]**

(c) One of Simon's results is anomalous.

(i) At which temperature did the anomalous result occur?

.... **[1]**

(ii) Suggest the correct value for the solubility at this temperature.

.... **[1]**

(d) (i) Describe what you would see if the solution Simon made at 70°C was cooled slowly to room temperature.

....

.... **[2]**

(ii) Explain your answer to (i).

....

.... **[2]**

(Total 11 marks)

3 Some coins are made of an alloy of zinc, nickel and copper. To find the percentage of zinc in the coins one coin, of mass 0.5 g, was placed in 25 cm^3 of hydrochloric acid of concentration 0.5 mol/dm^3. Only the zinc reacted.

$$Zn + 2HCl \rightarrow ZnCl_2 + H_2$$

When the reaction had finished the mixture was filtered and titrated against sodium hydroxide solution of 0.5 mol/dm^3 concentration.
To reach neutralisation point took 14.6 cm^3 of this sodium hydroxide solution.

(a) (i) How could you see when the reaction between the zinc and hydrochloric acid had finished?

.... **[1]**

Leave blank

(ii) Explain why the zinc reacted with the hydrochloric acid, but the nickel and copper did not.

....

.... **[1]**

(b) (i) Calculate the volume of 0.5 mol/dm³ hydrochloric acid which reacted with 14.6 cm³ of 0.5 mol/dm³ sodium hydroxide solution.

volume = ... cm³ **[2]**

(ii) Calculate the volume of 0.5 mol/dm³ hydrochloric acid which reacted with the zinc.

volume = ... cm³ **[1]**

(iii) Calculate the mass of zinc which reacts with this volume of 0.5 mol/dm³ hydrochloric acid.
(Relative atomic mass: Zn = 65)

mass = ... g **[3]**

(iv) What percentage of zinc was in the coins?

Leave blank

percentage = ... % **[2]**

(Total 10 marks)

4 The apparatus shown in the diagram was used to electroplate an iron medallion of mass 10.50 g with a metal M.
The metal M has a relative atomic mass of 59.

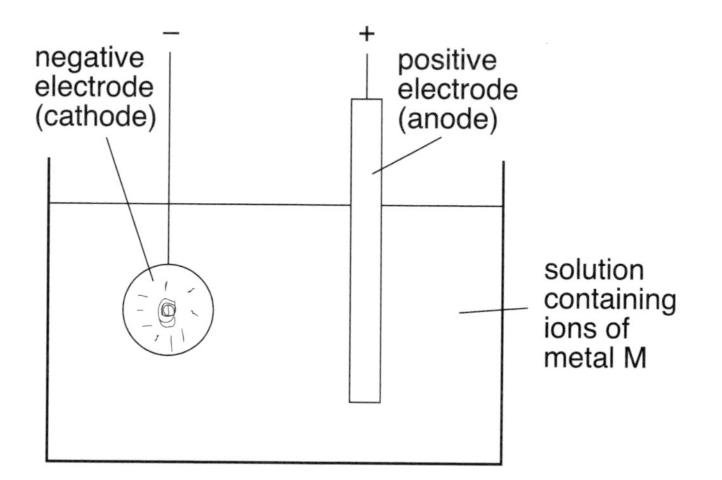

A current of 2 amperes was passed for 12 minutes.
After electroplating the medallion had a mass of 10.94 g.

(a) (i) Suggest one advantage there could be in electroplating the iron medallion.

....

.... **[1]**

(ii) Suggest a suitable material for the positive electrode (anode) in the apparatus.

.... **[1]**

Leave blank

(iii) Explain your choice in (ii).

....

....

.... **[1]**

(b) Calculate the following quantities. (1 Faraday = 96 500 coulombs.)

(i) The mass of M electroplated onto the medallion.

mass of M = ... g **[1]**

(ii) The quantity of electricity, in coulombs, passed through the circuit during the electroplating.

quantity of electricity = ... C **[2]**

(iii) The number of positive charges on an ion of the metal M.

number of positive charges = ... **[2]**

(Total 8 marks)

Leave blank

5 The table gives information about the atomic structure of some elements.

element	protons	arrangement of electrons	Group in the Periodic Table	metal or non-metal
Q	8	2, 6	6	....
X	....	2, 8, 1	1	metal
Y	17	....	7	non-metal
Z	18	2, 8, 8	....	non-metal

(a) Complete the table by filling in the four blank boxes. **[4]**

(b) Element X will react with another of the elements in the table to form a crystalline salt.
Write down the letter of this other element.

.... **[1]**

(c) Element X forms a compound with element Q.
Use the letters X and Q to show the formula of this compound.

.... **[1]**

(d) Element X will conduct electricity.
Use your knowledge of the structure of metals to explain this property.
Use a diagram to help your answer.

....

....

.... **[3]**

(e) Element Y is diatomic. Explain what this means.

....

.... **[2]**

(Total 11 marks)

Leave blank

6 Water hardness can be shown as the concentration of calcium ions in solution. The table shows the calcium ion concentration of samples of water from three different sources before and after boiling or addition of sodium carbonate.

		calcium ion concentration parts per million	
source	**treatment**	**before treatment**	**after treatment**
A	boiled	27	15
A	sodium carbonate	27	0
B	boiled	23	0
B	sodium carbonate	23	0

(a) **(i)** Explain how boiling removes hardness from water.
(One mark is for the correct use of scientific language.)

.... **[3+1]**

(ii) Boiling removed all of the hardness from the water from source B, but only part of the hardness from the water from source A.
Suggest a reason for this.

.... **[1]**

(b) The addition of sodium carbonate removed all of the hardness from the water from both sources.

(i) Explain how sodium carbonate removes hardness from water.
(One mark is for correct spelling, punctuation and grammar.)

.... **[3+1]**

Leave blank

(ii) Explain why using sodium carbonate removed all of the hardness from the water from both sources.

....

.... [1]

(c) Suggest a problem which might be caused by water hardness in the homes of people using water from source A.

....

....

.... [1]

(Total 11 marks)

Letts

Answers: GCSE Chemistry paper 1

Question			Answer	Mark
1	a	i	copper(II) sulphate solution	1
		ii	copper(II) oxide	1
		iii	$CuO + H_2SO_4 \rightarrow CuSO_4 + H_2O$	
			left side	1
			right side	1
		iv	neutralisation	1

Examiner's Tip
Neutralisation reactions occur not only between acids and alkalis, but also between metal oxides and acids, as in this example, and between carbonates and acids. A salt is always formed, in this case copper(II) sulphate.

Question			Answer	Mark
	b		Filter or decant off the clear blue liquid.	1
			Heat the solution to evaporate off some of the water.	1
			Leave the remaining solution to cool.	1
			+ 1 mark for logical order in answer	1

Examiner's Tip
Crystals will form if a hot saturated solution is allowed to cool to room temperature. The excess copper(II) oxide must be removed first, then some of the water evaporated to form a saturated solution of the copper(II) sulphate.
To score the extra mark you must make three points in the correct order.

Question			Answer	Mark
	c	i	exothermic	1
		ii		

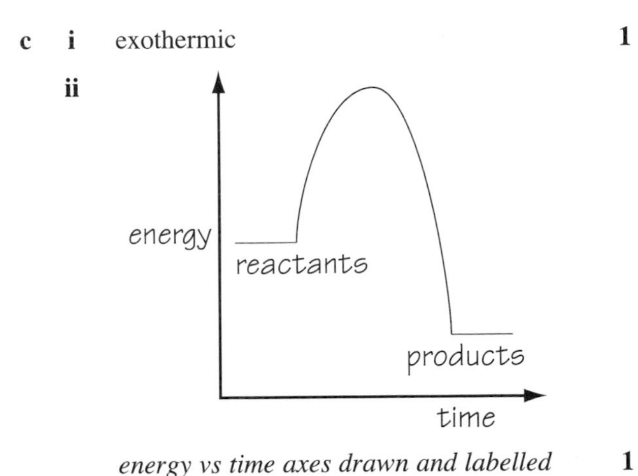

Answer	Mark
energy vs time axes drawn and labelled	1
reactants labelled at higher energy level than products	1
curve drawn to show progress of reaction	1

Examiner's Tip
Since an exothermic reaction gives out heat energy, the reactants must be at a higher energy level than the products. This is shown on the diagram.

Question			Answer	Mark
2	a	i	water	1
		ii	add to anhydrous copper(II) sulphate/cobalt chloride paper	1
			colour changes from white to blue/blue to pink	1

Examiner's Tip
The best test for water is to show that the boiling point of the liquid is 100 °C, but there is not enough water in this example to do this. Anhydrous copper(II) sulphate has lost its water of crystallisation, and is white. The water restores this water of crystallisation, returning the blue colour. Always describe how to do the test, and give the colour before and after the test.

Question			Answer	Mark
	b	i	white precipitate/white cloudiness/lime water turns milky	2
		ii	The gases contain carbon dioxide.	1

Examiner's Tip
All hydrocarbons burn to give water and carbon dioxide. Lime water forms a white solid of calcium carbonate when carbon dioxide is bubbled through it. This turns the solution cloudy – a white precipitate.

Question			Answer	Mark
	c	i	$2C_8H_{18} + 25O_2 \rightarrow 18H_2O + 16CO_2$	
			formulae	1
			balance	1

Examiner's Tip
This is a hard equation to balance. Two molecules of octane are needed in the equation so that an even number of oxygen atoms is used. Otherwise a half molecule of oxygen would be needed. Don't be afraid of large numbers of molecules in equations – sometimes they are necessary.

Question			Answer	Mark
		ii	When the octane does not have sufficient oxygen for complete combustion	1
			some of the carbon in the octane does not combine with oxygen.	1

Examiner's Tip
Hydrocarbons only burn completely to water and carbon dioxide if there is plenty of oxygen available. In air there is not enough oxygen, so the octane does not burn completely. All of the hydrogen forms water, but some of the carbon will form carbon monoxide or carbon. The carbon gives a sooty deposit.

Question	Answer	Mark
3 a		

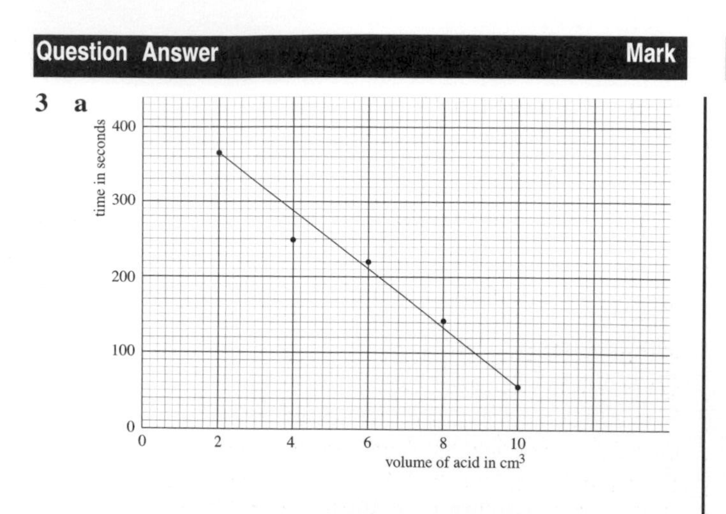

axes correctly drawn and labelled, including units 1
all points plotted to + or − half a square 1
a best fit line drawn ignoring the second point 1

Examiner's Tip
Axes need to be sensibly scaled and labelled with the thing being plotted, e.g. volume of acid, and the units, e.g. cm³. There is just one mark for doing this correctly for both axes! Plotting of the points must be accurate. Mark them clearly with a circle or cross. The best fit line must ignore any anomalous results.

b i This kept the total volume the same in each case, 1
otherwise the concentration would not have been proportional to the volume of acid added. 1

ii temperature (of the acid and water mixture)/stirring 1

Examiner's Tip
By using different volumes of acid diluted with water to the same total volume each time, Sarah made sure that the concentration was proportional to the volume of acid added. This could then be plotted to give the graph. Since the rate of a reaction increases with increase in temperature, this has to be kept constant if the investigation of rate with concentration of acid is to be a fair test.

c i 4 cm³ of acid 1

ii incorrect measurement of volumes or time/not constant temperature/inconsistent stirring/inconsistent tablets 1

Examiner's Tip
The result for 4 cm³ obviously does not fit onto a straight line, which this graph should have. There are many possible reasons for this, and any sensible suggestion would score the mark in (ii).

Question	Answer	Mark

d i Rate increases with increase in concentration. 1
Rate is directly proportional to concentration. 1

ii In order to react the acid particles need to collide with the solid sodium carbonate in the tablet. 1
At higher concentration there are more particles of acid per cm³, 1
therefore more particles collide with the sodium carbonate each second. 1

Examiner's Tip
As in many questions, the number of marks indicated for each part must be carefully noted. In both (i) and (ii) it would be easy to write less than the number of points needed for full marks. The rate of a reaction depends on the number of particles which collide each second. Of these collisions a proportion will result in the formation of products. The same proportion of a larger number of collisions will result in the formation of more product in a certain length of time, i.e. a greater rate of reaction.

4 a i mass of copper = 15.52 − 12.64 = 2.88 g 1

ii mass of oxygen = 15.88 − 15.52 = 0.36 g 1

b moles of copper = $\frac{2.88}{64}$ = 0.045 1

moles of oxygen = $\frac{0.36}{16}$ = 0.0225 1

mole ration of copper to oxygen = 0.045:0.0225 = 2:1 1

formula must be Cu_2O 1

Examiner's Tip
The masses of copper and oxygen are easily worked out by subtraction, but be careful that you are subtracting the correct figures! The mass of each element in the copper oxide is used to work out the number of moles by dividing mass by atomic mass. The number of moles of each element then gives the mole ratio. The tricky bit is converting this to a whole number ratio. Simply divide the larger value by the smaller one. The whole number ratio gives the number of atoms of each element: two for copper and one for oxygen.

c Sodium is much more reactive than copper, 1
so it is not so easy to remove the oxygen from sodium oxide/hydrogen will not remove the oxygen from sodium oxide. 1

Examiner's Tip
The more reactive a metal, the more energy is needed to pull oxygen away from the metal in the oxide. Hydrogen is not a strong enough reducing agent to remove the oxygen from sodium oxide.

Question Answer Mark

5 a i catalytic 1
cracking 1

Examiner's Tip
It would be very easy to answer 'cracking' for this question and get only one mark. Do not forget to look at the number of marks to be given for the answer and make sure you write a separate point for each one.

ii catalyst 1

iii small pieces have greater surface area 1
giving more collisions with the hydrocarbon per second 1

Examiner's Tip
The aluminium oxide is a solid catalyst for a reaction in which a gas is reacting. The amount of contact between the catalyst and the gas will have a large effect on how fast the reaction will go. Don't forget to use the ideas of time when you are writing about rate of reaction.

b i ethene 1

ii add bromine water 1
changes from red/orange to colourless 1

Examiner's Tip
The gas must be an alkene, and ethene is the alkene with two carbon atoms. Bromine water is the standard test for an alkene. The bromine reacts with the alkene and therefore is used up, leaving no colour in the mixture. An alkane will not decolourise bromine water.

c

```
  H H H H H H H H H H
  | | | | | | | | | |
H-C-C-C-C-C-C-C-C-C-C-H →
  | | | | | | | | | |
  H H H H H H H H H H

  H H H H H H H H          H     H
  | | | | | | | |           \   /
H-C-C-C-C-C-C-C-C-H  +       C=C
  | | | | | | | |           /   \
  H H H H H H H H          H     H
```

one mark for each graphical formula 3

Examiner's Tip
Remember to show each atom and each bond in a graphical formula, and don't forget the double bond in ethene. It is easy to give the wrong number of carbon atoms when there are so many, so count them to make sure they are correct for decane and octane.

Question Answer Mark

6 a i sandstone A 1

Examiner's Tip
Youngest rocks are always at the top of a sequence.

ii basalt 1

Examiner's Tip
Quite a tough question – the two igneous rocks that you have to know about for GCSE are basalt and granite. Questions are often about granite, shown as an igneous intrusion, but this is about basalt as a lava flow.

iii limestone 1

Examiner's Tip
Learn that limestone, marble and chalk are all mostly calcium carbonate.

b Since it contains crystals it is likely to be an igneous rock/formed from lava solidifying; since it contains very small crystals it must have cooled quickly/above ground. 2

Examiner's Tip
If you see two marks in a question like this, don't just put 'quick cooling' – think what else you could say.

c Fault X–Y has affected the mudstone layer but fault A–B has not; since the younger layers are on top of the older layers this must mean that X–Y occurred after A–B. 2

Examiner's Tip
In this sort of question, think about the order in which events must have taken place over geological time. Which layer was deposited first? Then what was next? When did fault A–B occur? You will then realise that X–Y must have taken place later than fault A–B.

7 a i respiration 1

ii $6CO_2 + 6H_2O$ (+ energy from sunlight) → $C_6H_{12}O_6 + 6O_2$
formulae 1
balance 1

iii from the sun/sunlight 2

Examiner's Tip
It is easy to get respiration and photosynthesis mixed up on the carbon cycle. Remember that respiration happens in both

Question	Answer	Mark

plants and animals, and produces CO_2, but photosynthesis happens in plants only, and uses up CO_2. You will get both marks even if the energy is not included in the equation in (ii). In (iii) you will not get the mark for 'light'.

---

b i As carbon dioxide was removed from the air by photosynthesis, 1
an equal amount 1
was returned to the air by respiration. 1

Examiner's Tip
The important idea here is that the removal and return of carbon dioxide was balanced, so that the percentage in the air remained constant.

---

ii carbon dioxide percentage has increased 1
caused by an increase in the burning of fossil fuels/
destruction of rain forests 1

c Hydrogen and helium have low densities. 1
They escaped from the Earth's atmosphere. 1
Ammonia reacted with oxygen to produce nitrogen and water 1
Bacteria (nitrifying or dinitrifying) removed ammonia. 1
+ 1 mark for correct spelling, punctuation and grammar 1
(Choose the best sentence and look for capital letters, spelling and absence of grammatical errors.)

8 a reversible/can go both ways/can form an equilibrium 1

Examiner's Tip
Any of the above answers are acceptable although saying that it shows that the reaction is reversible is probably the easiest to remember.

---

b nitrogen from air 1
hydrogen from crude oil/natural gas 1

c Increasing the pressure increases the rate of reaction (because there are a greater number of successful collisions between the greater numbers of particles present);
increasing the pressure increases the yield (because it pushes the equilibrium to the right and produces more ammonia). 2

Examiner's Tip
This is a difficult A* question. You have to use your knowledge of reaction rates to realise that increasing the pressure of gases increases their concentration and so will

Question	Answer	Mark

mean that more successful collisions will take place because there are more particles present. You also have to understand that since this equilibrium equation shows fewer molecules on the right hand side of the equation, increasing the pressure will force the equilibrium to the side with the smaller number of molecules, in this case to the right. This is often referred to as 'Le Chatelier's Principle' but knowing this name is not part of every examination board's specification.

---

d *Three from:*
Although the yield is high at low temperatures the rate of reaction is slow;
this is because gas particles have less energy and so there are fewer successful collisions;
using a higher temperature will give a lower equilibrium yield but this lower yield will be obtained much more quickly;
it is important economically to produce ammonia quickly and so higher temperatures are used which give a fast reaction rate;
a catalyst can be used to help speed up the rate at which the equilibrium is established. 3

Examiner's Tip
Another difficult question but the mark scheme enables you to score three marks without giving every single answer on the list. The examiner will look to see that you have understood the idea that too low a temperature will give too slow a rate of reaction because of fewer successful collisions between particles.

---

e Ammonia is used to make artificial fertilisers; 1
this was important because of the rapidly expanding world population requiring increased food production. 1

f $2 \times (14 + 3)$ 1
= 34 tonnes 1

9 a atomic number increases from Na – Cl – Ar/increases 11 – 17 – 18 1
number of electrons in outer shell increases: Na – 1, Cl – 7 and Ar – 8 1

Examiner's Tip
It is important to write about the number of protons, since this decides the element's position in the Periodic Table, and the number of electrons in the outer shell, since this determines the chemical behaviour of the element.

---

b Sodium is a very reactive metal. 1
Chlorine is a very reactive non-metal. 1
Argon is an unreactive gas. 1

Question	Answer	Mark
c	Sodium has one electron in its outer shell which is easily lost to get the stable electronic structure of argon – a typical metal property.	1
	Chlorine has seven electrons in its outer shell and easily gains one more to get the stable electronic structure of argon – a typical non-metal property.	1

Examiner's Tip
These two questions look at the relationship between the number of electrons in the outer shell of an atom and its chemical properties. The fact that chemical bonding leads to each atom having a full outer electron shell, which is the same electronic structure as a noble gas, is an essential feature of chemistry.

Question	Answer	Mark
10 a	*Either*	
	Calcium sulphate has a low solubility in water	1
	Calcium sulphate is precipitated	1
	or	
	Sea creatures with shells (crustaceans) use calcium ions to build up shells	1
	Shells are calcium carbonate	1
b	*Either*	
	$Ca^{2+}(aq) + SO_4^{2-}(aq) \rightarrow CaSO_4(s)$	2
	One mark for left hand side and one mark for right hand side.	
	or	
	$Ca^{2+}(aq) + CO_3^{2-}(aq) \rightarrow CaCO_3(s)$	2
	One mark for left hand side and one mark for right hand side.	

Examiner's tip
New specifications have questions on compositions of the oceans. There are two correct answers to this question. You will not be penalised if you miss out state symbols.
The table helps you with the correct formulae of calcium and sulphate ions.

Answers: GCSE Chemistry paper 2

Question	Answer	Mark
1 a		

name	formula	relative molecular mass	boiling point in°C	Mark
methane	CH_4	16	−161	
ethane	C_2H_6	30	……−88°C*….	1
..propane..	C_3H_8	44	−42	1
butane	….C_4H_{10}…	58	−1	1
pentane	C_5H_{12}	72	36	

* *(any negative value between −70 and −130 accepted)*

Examiner's Tip
Look carefully at the information in the table and use it to help fill in the blanks. You will see that each set of information has a pattern. Use the pattern to work out the correct value for the blank box.

Question	Answer	Mark
b	A homologous series is a series of compounds each differing from the last by the same group of atoms or each having the same general formula.	1
	Each alkane has CH_2 more than the one before or the general formula C_nH_{2n+2}.	1

Examiner's Tip
The question says 'as it applies to the alkanes' so your answer must say exactly how the term does apply to the alkanes.

Question	Answer	Mark
c i	Structural isomers have the same molecular formula	1
	but different structural (displayed) formulae.	1

Question	Answer	Mark
ii	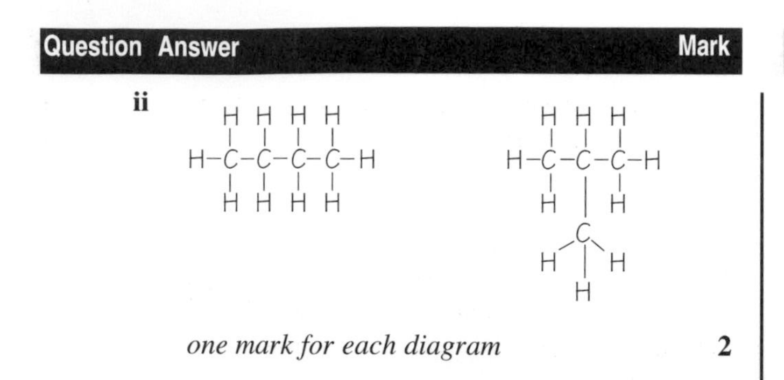	
	one mark for each diagram	2

Examiner's Tip
It is easy to draw a straight chain alkane with one of the carbon atoms pointing up or down and think this is an isomer. To be different the structural formula must actually have a carbon atom joined on in a different place.

2 a

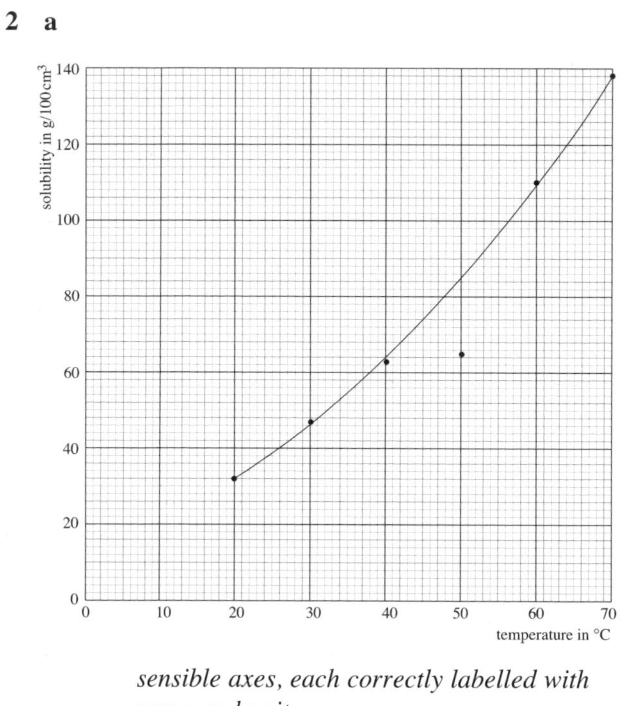

Question	Answer	Mark
	sensible axes, each correctly labelled with name and unit	1
	all points plotted correctly	1
	line passing through all points except the one for 50°C	1

Examiner's Tip
The value for 50°C is obviously incorrect and should be ignored when drawing the best fit line.

Question	Answer	Mark
b	Solubility increases with increase in temperature	1
	but is not directly proportional/increase becomes greater with higher temperature.	1

Examiner's Tip
You would not get a mark for saying 'solubility increases with temperature'. It increases with *increase* in temperature.

Question	Answer	Mark
c i	50°C	1
ii	86 g per 100 cm^3 water (+ or − 2 g)	1

Examiner's Tip
You can read off the correct value for 50°C on your graph. There is some allowance for your curve not being exactly the same shape as the Examiner's.

Question	Answer	Mark
d i	Crystals would appear	1
	and grow bigger.	1
ii	The solution at 70°C would be saturated with potassium nitrate.	1
	This would come out of solution as it cooled, so making crystals 'grow'.	1

Examiner's Tip
Note that there are two marks for each of these answers. Always look carefully at the number of marks and make sure you write one point for each mark.
For example, writing just 'crystals appear' for (i) would only score one mark.

Question	Answer	Mark
3 a i	no more bubbles/effervescence/hydrogen given off	1
ii	Zinc is higher in the reactivity series/ more reactive.	1
b i	$HCl + NaOH \rightarrow NaCl + H_2O$	
	mole ratio is 1 mole NaOH reacting with 1 mole HCl	1
	so 14.6 cm^3 0.5 mol/dm^3 NaOH reacts with 14.6 cm^3 0.5 mol/dm^3 HCl	1

Examiner's Tip
It is important to write the equation between hydrochloric acid and sodium hydroxide to find out that the mole ratio is 1:1. If you simply assume this you will lose a mark. Since the two solutions have the same concentration they will react in equal volumes.

Question	Answer	Mark
ii	volume of 0.5 mol/dm^3 HCl reacting with zinc = 25 − 14.6 = 10.4 cm^3	1

Examiner's Tip
This is a simple subtraction of the volume of acid used from the volume used originally to react with the zinc in the coin.

Question	Answer	Mark
iii	mole ratio from equation is 1 mole Zn to 2 moles HCl $10.4\,cm^3$ $0.5\,mol/dm^3$ HCl contains $0.5 \times \frac{10.4}{1000} = 0.0052$ moles	1
	moles Zn reacted = 0.5×0.0052 = 0.0026 moles	1
	mass Zn reacted = $0.0026 \times 65 = 0.169$ g	1

Examiner's Tip

You need to look back to the equation to see what the mole ratio of zinc to hydrochloric acid is. It is 1:2, so if you used 1:1 by mistake you would lose one mark. The volume of HCl can be used to calculate the moles of HCl, which then must be halved to get the moles of Zn.
Finally the moles of Zn must be multiplied by the relative atomic mass of zinc to get the mass in g.

Question	Answer	Mark
iv	% zinc in coins = $100 \times \frac{0.169}{0.5}$	1
	= 33.8 %	1

Examiner's Tip

The final stage is simply to divide the mass of zinc by the mass of the coin and multiply by 100 to get the %. Many candidates forget to multiply by 100.

Question	Answer	Mark
4 a i	stops rusting/stops corrosion/looks more attractive	1

Examiner's Tip

Most electroplating is carried out either to protect the original from corrosion or to improve the appearance of the product – often to make it look more expensive.

Question	Answer	Mark
ii	metal M/carbon/platinum	1
iii	the anode needs to be of the same metal as the electroplating to replace the ions removed from solution/the anode needs to be inert	1

Examiner's Tip

The answer to (iii) must match your answer to (ii). If a metal was used for the anode which then reacted in the electrolysis, e.g. copper, this would put ions of this second metal into the solution and spoil the electroplating.

Question	Answer	Mark
b i	mass of M = 10.94 − 10.50 = 0.44(g)	1

Examiner's Tip

This is simply a subtraction of the mass before electroplating from the mass after electroplating.

Question	Answer	Mark
ii	quantity of electricity = $2 \times 12 \times 60$	1
	= 1440(c)	1

Examiner's Tip

The amount of electricity used is calculated using the formula Q = It, in other words: coulombs = amps × minutes × 60. Note that time is in seconds for this calculation, hence the × 60. The answer is in coulombs of electricity.

Question	Answer	Mark
iii	1 mole of M would need $1440 \times \frac{59}{0.44} = 193091$ coulombs;	1
	this is 2 Faradays, so the number of positive charges on an ion of M = 2.	1

Examiner's Tip

The relative atomic mass of M is given in the question as 59. The answers from parts (i) and (ii) can be used to calculate the number of coulombs needed to deposit 1 mole of M. Since this value is 2 × 96500, i.e. 2 Faradays, the charge on the ion of M must be 2+. You need to set out your answer carefully, showing each calculation.

5 a

element	protons	arrangement of electrons	Group in the Periodic Table	metal or non-metal	Mark
Q	8	2, 6	6	non-metal	1
X	11	2, 8, 1	1	metal	1
Y	17	2, 8, 7	7	non-metal	1
Z	18	2, 8, 8	0	non-metal	1

Examiner's Tip

These answers can easily be worked out by using information already in the table, e.g. X has the electronic arrangement 2, 8, 1; this adds up to 11 electrons, so the atom must also have 11 protons.

Question	Answer	Mark
b	Y	1

Examiner's Tip

A crystalline salt can be formed from a metal atom from Group 1 and a non-metal atom from Group 7 joined together to make an ionic compound. X is a metal in Group 1 and Y is a non-metal in Group 7.

Question	Answer	Mark
c	X_2Q	1

Examiner's Tip

An atom of X has one electron in its outer shell to donate, and an atom of Q needs two electrons to join its outer shell for it to be full with 8. Two atoms of X will give one electron each to one atom of Q to form an ionic compound.

Question	Answer	Mark

The formula must therefore contain two atoms of X and one atom of Q.

d

Key
metal ion
electron

diagram similar to above 1
electrons between metal ions are mobile 1
and flow when pd is applied to give current through metal 1

Examiner's Tip
The diagram should clearly show metal ions with a 'sea' of electrons between them. It is a good idea to put a key next to the diagram. The 'sea' of electrons is responsible for the conduction of electricity through the metal. When a potential difference is applied these electrons move though the metal.

e Element Y is made of molecules 1
each containing two atoms. 1

6 a i Calcium hydrogencarbonate decomposes on heating. 1
Calcium carbonate is deposited as a solid. 1
This removes the calcium ions causing hardness from solution. 1
+ 1 mark for correct use of scientific words 'decompose' and 'hardness' 1

ii The water from source A must contain permanent hardness
(as well as temporary hardness). 1

Question	Answer	Mark

Examiner's Tip
Temporary water hardness is caused by calcium hydrogen carbonate dissolved in the water. When this is decomposed by heat it forms insoluble calcium carbonate which is precipitated as a white solid. Since the calcium ions responsible for the water hardness are now in the solid calcium carbonate, not in solution, the water is soft. Permanent water hardness is caused by calcium sulphate, which is not decomposed by heat and therefore causes the water still to be hard after boiling.

b i A precipitate of calcium carbonate is formed 1
removing the calcium ions from solution 1
$Ca^{2+}(ag) + CO_3^{2-}(ag) \rightarrow CaCO_3(s)$ 1
+ 1 mark for correct spelling, punctuation and grammar. 1

ii All calcium ions are removed when sodium carbonate is used. 1

Examiner's Tip
Since calcium carbonate is insoluble in water, adding carbonate ions, in the sodium carbonate, to the hard water will cause calcium carbonate to form a precipitate. Without calcium ions the water is soft. It does not matter whether the calcium ions are from permanent or temporary hardness, they will all be removed from solution.

c scum formed with soap/kettles fur up 1

Examiner's Tip
Calcium ions form an insoluble precipitate, called scum, with soap. This means that more soap has to be used, and the scum can stain clothing. If temporary hardness is present in water boiled in a kettle, the calcium carbonate formed collects in the kettle; this is called furring up. It can cause failure of the element in an electric kettle or washing machine.

Grade Predictor
The grid below suggests the grades that you might expect to achieve on these papers. Remember that the grade boundaries change from year to year at each Examination Board.

A*	135–160
A	110–134
B	90–109
C	70–89
D	45–69

Index

numbers in italics refer to diagrams: numbers in bold refer to tables